Giving Voice

The John D. and Catherine T. MacArthur Foundation Series on Digital Media and Learning

Civic Life Online: Learning How Digital Media Can Engage Youth, edited by W. Lance Bennett

Digital Media, Youth, and Credibility, edited by Miriam J. Metzger and Andrew J. Flanagin

Digital Youth, Innovation, and the Unexpected, edited by Tara McPherson

The Ecology of Games: Connecting Youth, Games, and Learning, edited by Katie Salen

Learning Race and Ethnicity: Youth and Digital Media, edited by Anna Everett

Youth, Identity, and Digital Media, edited by David Buckingham

Engineering Play: A Cultural History of Children's Software, by Mizuko Ito

Hanging Out, Messing Around, and Geeking Out: Kids Living and Learning with New Media, by Mizuko Ito et al.

The Civic Web: Young People, the Internet, and Civic Participation, by Shakuntala Banaji and David Buckingham

Connected Play: Tweens in a Virtual World, by Yasmin B. Kafai and Deborah A. Fields

The Digital Youth Network: Cultivating Digital Media Citizenship in Urban Communities, edited by Brigid Barron, Kimberley Gomez, Nichole Pinkard, and Caitlin K. Martin

The Interconnections Collection, developed by Kylie Peppler, Melissa Gresalfi, Katie Salen Tekinbaş, and Rafi Santo

> *Gaming the System: Designing with Gamestar Mechanic*, by Katie Salen Tekinbaş, Melissa Gresalfi, Kylie Peppler, and Rafi Santo
>
> *Script Changers: Digital Storytelling with Scratch*, by Kylie Peppler, Rafi Santo, Melissa Gresalfi, and Katie Salen Tekinbaş
>
> *Short Circuits: Crafting E-Puppets with DIY Electronics*, by Kylie Peppler, Katie Salen Tekinbaş, Melissa Gresalfi, and Rafi Santo
>
> *Soft Circuits: Crafting E-Fashion with DIY Electronics*, by Kylie Peppler, Melissa Gresalfi, Katie Salen Tekinbaş, and Rafi Santo

Connected Code: Children as the Programmers, Designers, and Makers for the 21st Century, by Yasmin B. Kafai and Quinn Burke

Disconnected: Youth, New Media, and the Ethics Gap, by Carrie James

Education and Social Media: Toward a Digital Future, edited by Christine Greenhow, Julia Sonnevend, and Colin Agur

Framing Internet Safety: The Governance of Youth Online, by Nathan W. Fisk

Connected Gaming: What Making Video Games Can Teach Us about Learning and Literacy, by Yasmin B. Kafai and Quinn Burke

Giving Voice: Mobile Communication, Disability, and Inequality, by Meryl Alper

Giving Voice

Mobile Communication, Disability, and Inequality

Meryl Alper

The MIT Press
Cambridge, Massachusetts
London, England

This book was set in Stone Sans and Stone Serif by Toppan Best-set Premedia Limited. Printed and bound in the United States of America.

Library of Congress Cataloging-in-Publication Data

Names: Alper, Meryl, author.
Title: Giving voice : mobile communication, disability, and inequality / Meryl Alper.
Description: Cambridge, MA : MIT Press, [2017] | Series: The John D. and Catherine T. MacArthur Foundation series on digital media and learning | Includes bibliographical references and index.
Identifiers: LCCN 2016022991 | ISBN 9780262035583 (hardcover : alk. paper) | ISBN 9780262533973 (pbk. : alk. paper)
Subjects: LCSH: Communication devices for people with disabilities--Social aspects. | Voice output communication aids--Social aspects. | Assistive computer technology--Social aspects. | Sociology of disability.
Classification: LCC HV1568.4 .A47 2017 | DDC 362.4/0483--dc23 LC record available at https://lccn.loc.gov/2016022991

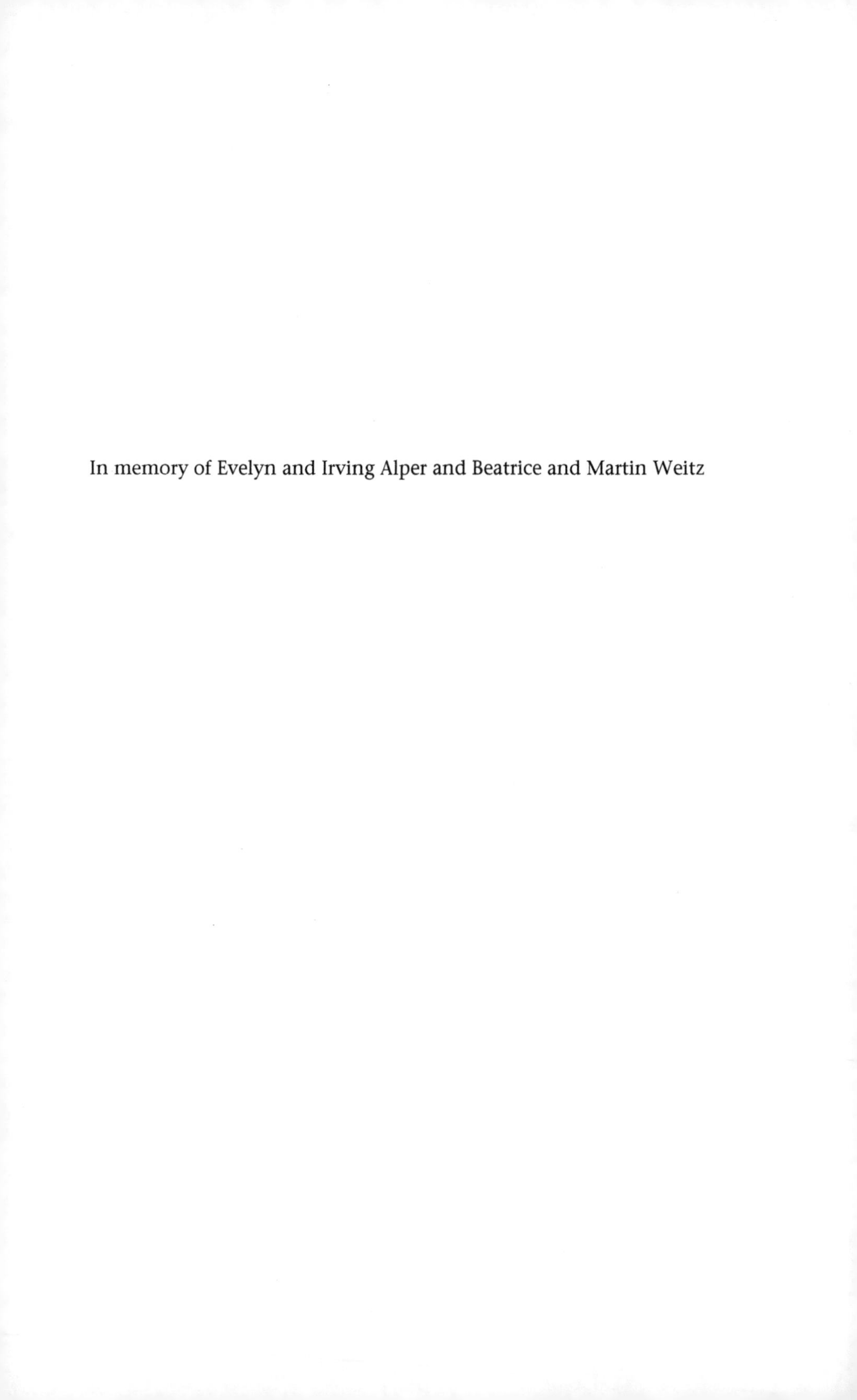

In memory of Evelyn and Irving Alper and Beatrice and Martin Weitz

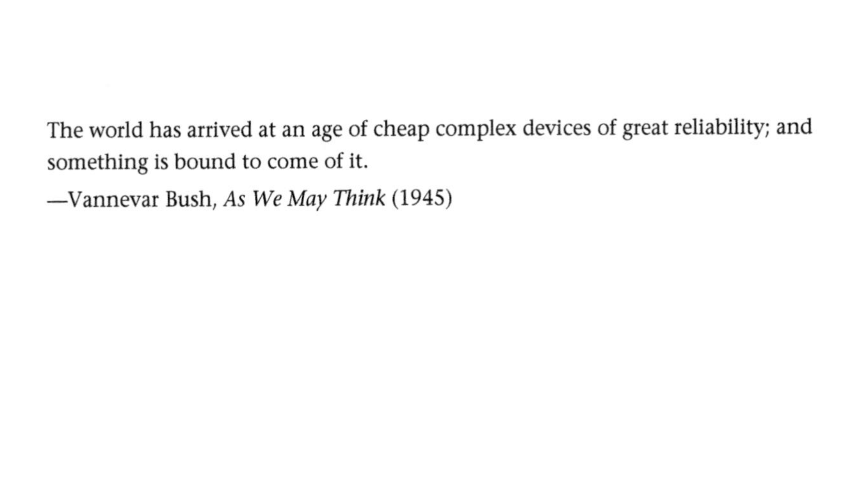

The world has arrived at an age of cheap complex devices of great reliability; and something is bound to come of it.

—Vannevar Bush, *As We May Think* (1945)

Contents

Series Foreword

In recent years, digital media and networks have become embedded in our everyday lives, and are part of broad-based changes to how we engage in knowledge production, communication, and creative expression. Unlike the early years in the development of computers and computer-based media, digital media are now *commonplace* and *pervasive*, having been taken up by a wide range of individuals and institutions in all walks of life. Digital media have escaped the boundaries of professional and formal practice, and the academic, governmental, and industry homes that initially fostered their development. Now, diverse populations and noninstitutionalized practices, including the peer activities of youths, have embraced them. Although specific forms of technology uptake are highly diverse, a generation is growing up in an era when digital media are part of the taken-for-granted social and cultural fabric of learning, play, and social communication.

This book series is founded on the working hypothesis that those immersed in new digital tools and networks are engaged in an unprecedented exploration of language, games, social interaction, problem solving, and self-directed activity that leads to diverse forms of learning. These diverse forms of learning are reflected in expressions of identity, how individuals express independence and creativity, and those individuals' ability to learn, exercise judgment, and think systematically.

The defining frame for this series is not a particular theoretical or disciplinary approach, nor is it a fixed set of topics. Rather, the series revolves around a constellation of topics investigated from multiple disciplinary and practical frames. The series as a whole looks at the relation between youth, learning, and digital media, but each contribution might deal with only a subset of this constellation. Erecting strict topical boundaries would exclude some of the most important work in the field. For example, restricting the content of the series only to people of a certain age would mean artificially reifying an age boundary when the phenomenon demands otherwise. This would become especially problematic with new forms of online participation where one crucial outcome is the mixing of participants of different ages. The same goes for digital media, which are increasingly inseparable from analog and earlier media forms.

The series responds to certain changes in our media ecology that have important implications for learning. Specifically, these changes involve new forms of media *literacy* and developments in the modes of media *participation*. Digital media are part of a convergence between interactive media (most notably gaming), online networks, and existing media forms. Navigating this media ecology involves a palette of literacies that are being defined through practice yet require more scholarly scrutiny before they can be fully incorporated pervasively into educational initiatives. Media literacy involves not only ways of understanding, interpreting, and critiquing media but also the means for creative and social expression, online search and navigation, and a host of new technical skills. The potential gap in literacies and participation skills creates new challenges for educators who struggle to bridge media engagement inside and outside the classroom.

The John D. and Catherine T. MacArthur Foundation Series on Digital Media and Learning, published by the MIT Press, aims to close these gaps, and provide innovative ways of thinking about and using new forms of knowledge production, communication, and creative expression.

Acknowledgments

Though my name alone appears on the cover of this book, which began as my dissertation, I have many people to thank for helping me see it through to the end.

First and foremost, I am endlessly appreciative of my loving family: my parents, Alyse and Andrew, and my sisters, Taryn and Devra. Their pride in me is life's greatest reward. I am the child of book lovers and parents who love through books, be it my mom's eagerness to add *Giving Voice* to the collection of our hometown New City Library, or my dad's enthusiasm for reading bedtime story after bedtime story nightly while growing up. Besides providing moral support, Taryn (a speech-language pathologist who works with autistic children) and Devra (an occupational therapist with a deep curiosity for assistive technology) also lent me their clinical expertise and professional perspectives.

I had the great privilege of working with an all-star committee at the Annenberg School for Communication and Journalism at the University of Southern California. My gratitude for Henry Jenkins—my doctoral adviser, dissertation chair, and personal Yoda—goes beyond words, except to say that "the force is strong with this one." This book also bears the indelible mark of Ellen Seiter, whose critically engaged work on children, parents, and media significantly influenced my own, and early interest in the project was profoundly motivating. I thank Mike Ananny for his carefully considered feedback, particularly his insights into the complex relationship between apparatuses for voice and listening. I am also deeply indebted to disability and mass media scholar Beth Haller, who offered unwavering support from afar and a steady stream of relevant news articles to read.

At the University of Southern California, Paul Lichterman's Qualitative Research Methods course in the sociology department jump-started this project and got me to enter the field. He and my fellow classmates provided a safe space to workshop my ideas, and refine my observation and interviewing skills. In the occupational therapy department, Rachel Proffitt's class on assistive technology exposed me to the wide world of augmentative and alternative communication as well as to various disability communities in the greater Los Angeles area.

Thank you to the Annenberg Foundation for generously supporting my five years of graduate education with an Annenberg Fellowship. The staff and administration at the Annenberg School also provided significant resources. I am especially appreciative of Sarah Banet-Weiser, Anne Marie Campion, Amanda Ford, G. Thomas Goodnight, Larry Gross, Christine Lloreda, Imre Meszaros, Peter Monge, and Billie Shotlow.

The MIT Press and John D. and Catherine T. MacArthur Foundation Series on Digital Media and Learning has been a perfect home for this project. My editor, Susan Buckley, was a pleasure to work with throughout the writing and revision process. Thank you to the anonymous reviewers of the manuscript for their thoughtful feedback. Lynn Schofield Clark, Vikki Katz, and Sarita Yardi Schoenebeck also generously read chapter drafts in the final stages of manuscript preparation.

I am grateful for my welcoming colleagues in the Department of Communication Studies at Northeastern University, where I began teaching in September 2015. Thank you in particular to Carole Bell, Dale Herbeck, Sarah J. Jackson, Joseph Reagle, and Brooke Foucault Welles. Thanks also to my undergraduate students for their curiosity and conversation.

Having conducted research in the area of youth and communication technology since my own undergraduate years at Northwestern University, I have had truly amazing opportunities to learn formally and informally from some of the best in the field. I was inspired to pursue a PhD after having been mentored professionally by three successful women with doctorates of their own. Thank you to Alisha Crawley-Davis, Jennifer Kotler, and Christine Ricci for modeling excellence during my time at Sesame Workshop and Nickelodeon, and their continued support and encouragement. A hearty thanks additionally to Sandra Calvert, Justine Cassell, Maya Götz, Nancy Jennings, Amy Jordan, David Kleeman, Dafna Lemish, Sonia Livingstone, Barbara O'Keefe, and Ellen Wartella for their collective wisdom and career guidance.

I came to disability research by way of an interest in diversifying the study of children, families, and media in the digital age. Over the years, Elizabeth Ellcessor, Katie Ellis, Shuli Gilutz, Gerard Goggin, Kevin Gotkin, Sara Hendren, Juan Pablo Hourcade, Mara Mills, and Melissa Morganlander have all been important conversation partners in the interdisciplinary study of disability, communication, and technology.

Parts of this project were presented and discussed at the Oxford Internet Institute Summer Doctoral Program, Berkman Center for Internet and Society at Harvard University, Social Media Collective at Microsoft Research New England, Department of Informatics at the University of California at Irvine, Digital Media and Learning Conference, iConference, and International Communication Association's annual conferences. I am grateful to the scholars who offered feedback in these venues as well as career and publishing advice, particularly Nancy Baym, danah boyd, Paul Dourish, Nathan Ensmenger, Tarleton Gillespie, Mary Gray, Gillian Hayes, Philip Howard, Mizuko Ito, Alice Marwick, Vikki Nash, Gina Neff, and Katy Pearce.

I am especially appreciative of the highly collaborative, socially supportive, and intellectually generous group of Annenbergers past and present whom I've had the opportunity to know and learn from over the years. This includes Alison Bryant, Kevin Driscoll, Laurel Felt, Becky Herr-Stephenson, Becca Johnson, Neta Kligler-Vilenchik, Alex Leavitt, Julien Mailland, Andrew Schrock, Lana Swartz, and Nikki Usher. Special shout-out to Kate Miltner for nourishing my spirit during the stressful home stretch of the PhD program and sustaining me with many dinners of snacks.

Over the course of my graduate studies I was lucky to build a robust support system among my fellow doctoral students in communication and related fields. Special thanks to my OII people—Stacy Blasiola, Caroline Jack, Jenna Jacobson, Stephanie Steinhardt, and Misha Teplitskiy—for our never-ending Facebook conversation thread. Thanks also to Morgan Ames, Dixie Ching, Sabrina Connell, Katie Day Good, Brian Keegan, J. Nathan Matias, Laine Nooney, Rebecca Onion, Cassidy Puckett, Rebekah Pure, Matt Rafalow, Joy Rankin, Ricarose Roque, and Rafi Santo.

I am deeply fortunate to have dear friends across the country, and especially in Los Angeles and New York, for both consolation and celebration. Much love to Dawn Amodeo, Ev Boyle, Erika Brooks Adickman, Darleen Chyu, Rebecca Eskreis, Eric Fingerman, Marnie Kaplan, Jeanne Leitenberg, Analise McNeill, Thuy-Van Nguyen, Erin Olson, Steve Pomerantz, Jackie Shlomi, Lisa Slopey, Jennifer Vecchiarello, and Erin Ward.

This project and book would not have been possible without those in the Los Angeles area who shared my recruitment materials with speech-language pathologists, educators, nonprofit groups, and parent e-mail lists. Many thanks to the staff members at the Evergreen Assistive Technology

Hub and Rossmore Regional Center—especially Rachel and Caren, who gave so generously of their time and energy.

And finally, my sincerest appreciation goes to those who contributed the most to this book: the parents and children who opened up their homes and lives to me. I hope that in some small part, this book reflects all you've taught me about the powerfully wordless ways to express love and gratitude.

1 Introduction

"For Children Who Cannot Speak, a True Voice via Technology," the headline read. In 2012, the *New York Times* profiled one such child, nine-year-old Enrique Mendez. He was born with the developmental disability Down syndrome and speech apraxia, a motor disorder in which the brain cannot coordinate the body parts needed to produce oral speech.[1] Enrique has difficulty saying sounds, syllables, and words, but he can better ask his brother to play with wrestling figurines and greet his parents with "I love you" in the morning when using an Apple iPad along with an app named Proloquo2Go. The app converts the icons and text that Enrique selects on the tablet's touch screen into synthetic speech output that his nearby conversation partners can hear.

Similar headlines about the iPad and Proloquo2Go, and the novel use of mobile technologies to generate voice, appear in newspapers like the *Boston Globe* and on television outlets such as CNN: "Technology Helps Autistic 12-Year-Old Find a Voice for His Bar Mitzvah" and "How Tablets Helped Unlock One Girl's Voice."[2] These contemporary reports are nearly indistinguishable from ones published decades earlier that also sang the praises of advancements in portable communication aids for individuals with communication disabilities. A 1977 *Wall Street Journal* article on a "hand-held device" known as the Phonic Mirror HandiVoice characterized it as "Offering an Electronic Voice to Vocally Impaired."[3] Two years later, a profile in the *Los Angeles Times* on a local area girl's use of another technology, the Canon Communicator, led with the headline "Electronic Help for the Handicapped: The Voiceless Break Their Silence."[4]

Today's sleek mobile communication technologies are completely unrecognizable compared with their clunky predecessors from the late 1970s and early 1980s, yet the headlines have not changed much. Each article mentioned above speaks volumes about the rhetoric of revolution embraced by technophiles, paternalistic discourses of technology as an equalizer of opportunity and access, and notions of voice as both symbolizing human speech and serving as a powerful metaphor for agency, authenticity, truth, and self-representation. All of the stories focus on objects, not people, when they frame mobile communication technologies as a medium for voice, tool for finding voice, and metaphoric key for freeing a caged voice. And none provide much insight into what does and does not get said through or about these speech-generating technologies in the long term after the journalists and users part ways.

This book is about what happens next. It centers on the social implications of communication technologies that purport to "give voice to the voiceless," and explores the varied meanings of this phrase through the critical lens of disability. For its part, AssistiveWare, the Dutch technology company that produces Proloquo2Go, says that the app "provides a 'voice'" to individuals with complex communication needs and also refers to the app *as* voice, with the tagline "A voice for those who cannot speak."[5]

In a rapidly changing media ecology and political environment, the question is not only which voices get to speak, but also who is thought to have a voice to speak with in the first place. The values, desires, and ideals of dominant cultural groups are systematically privileged over others in societies, and these biases are built into the very structures of organizations, institutions, and networks. "Giving voice to the voiceless" regularly stands in for the idea that the historically disadvantaged, underrepresented, or vulnerable gain opportunities to organize, increase their visibility, and express themselves by leveraging the affordances of information, media, and communication technologies.[6] Besides the iPad and Proloquo2Go, an endless list of technologies and platforms—including civic media, mobile Internet, Twitter, community radio, and open data—are all imbued with voice-giving qualities.[7]

These tools may selectively amplify voices within and across various publics and audiences, but their existence does not automatically call the status quo of structural inequality (i.e., racism, patriarchy, misogyny, and homophobia) into question. In particular, it is counterproductive when these same discourses about voice employ disability as a metaphor in the service of another's cause. More often than not, media and communication studies scholars call on disability as emblematic of the human experience, such as in Marshall McLuhan's "prosthesis" or Donna Haraway's "cyborg." When Nick Couldry writes that state actors can be "'voice blind,' that is, blind to the wider conditions needed to sustain new and effective forms of voice," this language evokes both speech and visual impairment without directly addressing the systemic violences that attempt to silence those with disabilities.[8]

Such instrumentalization underpins the message behind the slogan "Nothing about us without us," which US disability activists first took up in the 1990s to describe their demands for active involvement in the planning of policies affecting their lives and their disdain for external sanctioning to speak up.[9] One billion people, or 15 percent of the world's population, experience some form of disability, making them one of the largest (though also most heterogeneous) groups facing discrimination worldwide.[10] Efforts to better include individuals with disabilities within society through primarily technological interventions rarely take into account all the other ways in which culture, law, policy, and even technology itself can also marginalize and exclude.[11]

These conditions overlap in ways that both enable and disable societal, cultural, and civic participation for those with disabilities, revealing contradictions in the modern human experience. Enrique and other disabled youth serve as cultural symbols onto which the able-bodied project their hopes and anxieties about health, well-being, and the future.[12] Mass media tend to depict youth with disabilities as beneficiaries of technology, while hailing well-intentioned engineers, scientists, and technologists (often white, male, and able-bodied) as their benefactors. Such portrayals distract us from seeing children, adolescents, and teenagers with disabilities as young people whose experiences with communication technology can be ordinary and even mundane. They preclude researchers from asking nuanced questions about the social and cultural contexts of their technology use and non-use. And perhaps most important, they mask how other dimensions of difference besides disability—such as class, race, ethnicity, gender, sexuality, and nationality—shape how youth with disabilities consume, create, and circulate media.

Beyond the hype and hyperbole, I argue in the following pages that technologies largely thought to universally empower the "voiceless" are still subject to disempowering structural inequalities. Over sixteen months, I engaged in qualitative fieldwork with the families of twenty young people in the greater Los Angeles area. For inclusion in the study, each child was required to be between the ages of three and thirteen at the start of research, have a developmental disability (such as autism and cerebral palsy), be either unable to produce oral speech or have significant difficulty doing so, and use the iPad and Proloquo2Go as their primary mode of communication (for a more in-depth discussion, see the chapter on methods).[13] Parents came to obtain their child's iPad and Proloquo2Go through a variety of strategies, including direct purchase, charitable donation, temporary loan, and school provision.

Holding these participant parameters and communication system as constants, I observed children and their parents receiving at-home training on how to use the technology from two speech-language pathologists named Rachel and Caren, both of whom were under contract with a local disability resource center in Central Los Angeles.[14] I conducted in-depth interviews with some of these parents and others throughout Southern California (in Los Angeles, Orange, and Ventura counties). I interviewed assistive technology specialists in local school districts who frequently interact with children who use the iPad and Proloquo2Go, interface with their families, and come into contact with other actors who directly and indirectly influence reception of the technology, including insurance company representatives and support staff in the assistive technology industry.

Throughout this fieldwork, my research questions centered on how parents managed their child's use of the iPad and Proloquo2Go as well as other communication technologies, and how they incorporated media into their family's daily life at home, on the go, and in the community. All of the parents I spoke with, regardless of their circumstances, stressed that they wanted the best for their child. Many believed wholeheartedly that mobile media could be powerful tools for their children to more effectively communicate their needs, preferences, and desires, and assert more control over their sometimes-chaotic lives. Each one was actively trying to maintain dignity in a world all too quick to strip them of it.

Against this shared background and these efforts for digital equity, clear distinctions emerged as these families adapted around a new set of routines. To borrow the phrasing of media scholars Leah Lievrouw and Sonia Livingstone, the social meanings that parents derived from these technologies along with the social consequences of them differed subtly and not so subtly across class, with additional considerations for gender, race, and ethnicity.[15] Working-class and low-income parents talked about the iPad, Proloquo2Go, and other communication technologies in ways that were often out of sync with how middle- and upper-class school district staff, therapy providers, and the mass media characterized them. Whether intentionally or inadvertently, educational, medical, and media institutions preserved as well as reinforced the more privileged status of middle- and upper-class parents within this ecosystem, which directly and indirectly impacted less privileged lower-class families and the capacity to support their child.

Drawing on sociologist Pierre Bourdieu's theorization of capital and its application to research on education, parenting, and technology, I detail over the course of this book how parents' ability to mobilize social, economic, and cultural capital shaped the extent to which their children could not only speak but also be heard.[16] In short, physically handing someone a tablet that talks does not in and of itself give that person "a true voice." Nor, contrary to a legion of pop psychologists, are handheld mobile devices single-handedly disabling people's empathy and capacity for face-to-face communication.[17] Rather, voice is an overused and imprecise metaphor—one that abstracts, obscures, and oversimplifies the human experience of disability. Empirically investigating the use of mobile devices as synthetic speech aids provides a novel way of understanding voice and communication technologies. At its heart, this book supports the rights of all individuals in society to determine their own conceptual sense of voice, and to use those voices to feel known in the world.

Broken Records

Detailed below, Karun's experiences at opposite ends of the socioeconomic spectrum as a mother and immigrant to the United States offers a powerful entry point for understanding the complex role of privilege, social class, and capital in how US parents, including those of children with disabilities, support their children's communication skills and use of media and communication technologies.

"Back home, I used to play piano with him. Now I don't have piano," Karun said with a heavy heart. In the time and space between "back home" and "now," a civil war that broke out in Syria in spring 2011 had escalated, endangering her and her husband, Mihran, and their two sons, Pargev (age thirteen) and Joseph (age nine). At the time of writing this book, there are over four million Syrians refugees displaced globally due to the humanitarian crisis.[18] Since I interviewed Karun in fall 2013, more than half of all US governors have announced that they oppose settling Syrian refugees in their states, stoking the fires of xenophobia and Islamophobia following the revelation that one of the suspects in the Paris terrorist attacks of November 2015 was granted entry to Europe among a wave of refugees by using documents falsely identifying him as Syrian.

Karun described a life of relative privilege in Syria prior to the war. Both she and Mihran attended private schools where they learned English. Mihran had studied abroad in England and became a radiologist in Syria. "I had *plenty* of time over there," Karun remarked, describing life in her home country. "My housework was done by a nanny. I could afford there, a nanny. She used to help in cooking and cleaning the house." Karun also provided her children with various enrichment activities such as horseback riding, swimming lessons, and the aforementioned piano lessons.

"The war came very fast," Karun explained. She and her family sought asylum in California, where her relatives had settled years earlier, along with a large diasporic Armenian community. From 1980 to 2013, the Armenian population in Los Angeles County tripled to 170,000.[19] When I met Karun, both she and her husband were unemployed, and the family was living on temporary refugee support from the US government. Not only did Karun not have her piano, but "right now, I don't have time," she said. "That's the bad thing here in the United States. Life is stressful. ... It's just run, run, run, run." The abandoned piano was a metaphor for the loss of a privileged life in Syria and adoption of a new, lower-class and marginalized identity in the United States.

The piano, however, also symbolized another kind of longing. When Karun remarked, "I used to play piano with him," she referred specifically to Pargev, who is autistic and has significant difficulty speaking. Karun thought that practicing the piano would be a more worthwhile leisure activity than how Pargev currently spent his free time at home. "Instead of playing with water or stimulatory behaviors, I want him to do something functional," Karun said. During each of the three 1.5-hour visits that I made to the family's Los Angeles apartment, Pargev engaged in self-stimulatory behavior, also known more colloquially as stimming. Many autistic people report that the repetition of physical movements or movement of objects helps them maintain emotional balance, regulates their senses, and provides pleasure.[20] Everyone stims to some degree, perhaps fiddling with a bookmark as you read this book. Suspending my own judgment of Pargev's behavior, he seemed calm and content to pour food and beverages, like chips and soda, back and forth into plastic bowls and cups of uniform sizes, taking periodic bites and sips.

Besides the piano, many of the other resources that Karun accrued had to be left behind in Syria. She explained how a few years earlier, "I was saving some money either to buy an iPad, because they are like $1,000, or remodel my rooftop to make it a play area for Pargev." While she had heard that autistic children were benefiting from the iPad, she chose the play area as a longer-term investment. During the war, though, the rooftop became unsafe. She recalled, "All the time there's airplanes, the military airplanes. And plus lots of people were killed by just a bullet, just a running bullet, going through accidentally." Karun ultimately regretted her decision, saying, "I didn't know it was going to be ruined, and we're going to leave and come here. I wish I'd bought from those days, the iPad."

After immigrating to California, Karun managed to acquire the device through a charitable donation. She was "really hoping to find something useful for [Pargev] on the iPad." In lieu of a physical piano, she downloaded "this little piano game" onto the device. Pargev, however, was not interested. "I wish he loves games," Karun remarked wistfully. Unlike the dominant cultural figure of the mother who sees no value in video gaming, Karun characterized the activity (as well as piano playing) as "something functional"—a category to which stimming, according to her, did not belong. Along with the iPad, the charity provided Karun with a gift card to purchase the Proloquo2Go app. "Right now, he can say three-word sentences like, 'Give me please.' 'Move please,' 'I want juice.' Only three words, not more than three words," Karun said. She hoped that Proloquo2Go would expand Pargev's communication.

Karun wanted a better life for her son, but felt that she was getting little school support. Even though Karun had the donated iPad and Proloquo2Go, Pargev's school supplied its own copy of the hardware and software for him to use in the classroom with teacher and therapist supervision. While the school allowed Pargev to take its iPad back and forth between home and school, Karun had received little hands-on training on how best to use the technology to communicate with Pargev at home. "For this Proloquo, honestly, he needs [a] professional with me. I can do it, I can help him to use it constantly," she maintained, but the sporadic at-home training sessions she received from Rachel through California's Department of Developmental Services were "not enough." Insufficient family support from therapists and school staff as well as the lack of trained personnel are well-documented challenges for youth with disabilities and their families.[21]

"This is something that disappointed me in [the] United States," Karun commented. "They told me, 'Once you go to [the United States], you're going to be relieved and they take care of your child,' but it wasn't like that." In Syria, Karun had homeschooled Pargev. "Over there," she explained, "I'm in control. I can see what's going on." In the United States, she had less power over his learning, and felt that Pargev was regressing as a result. Karun noted that "Pargev knew the alphabet when he was four and a half. I used to contact with the teacher [in California] and tell her, 'Please teach him to write.' 'It's early,' she told me. 'It's early. It's early.' Always you get these answers." Karun had left her piano behind in Syria and tried her best to re-create it through a piano app on the iPad in the

United States; instead, she and Pargev wound up listening to a broken record of false hopes.

Cultural Capital

In her essay "Practicing at Home: Computers, Pianos, and Cultural Capital," Ellen Seiter draws an extended analogy between pianos and computers.[22] The piece relates to Karun's story not only because it involves those same technologies but also because it offers a relevant theoretical framework through which to understand the role of social class in family media use. Seiter explains that baby grand pianos and personal computers have each historically served as an "instrument of modern education" in upper- and middle-class US homes.[23] Privileged children tend to gain more experience with these learning machines, and earlier in life, than do working-class children. They learn specific "codes" at home that less well-off children do not, be it computer keyboarding or playing the piano keys.[24] Educational institutions systematically reward students who can demonstrate the kinds of technological proclivities and literacies that upper- and middle-class children are more likely to have acquired outside school.[25] The higher status that schools associate with these seemingly "natural" competencies leads to the reproduction of social inequality, or what Seiter terms the "home technology divide."[26]

This gap persists not only due to household-level economic disparities but also parents' unequal access to capital, including the social and cultural resources that they gain through their own education, careers, neighborhoods, friends, and extended family.[27] Bourdieu theorized that three main forms of capital—economic, social, and cultural—structure our world.[28] Economic capital is the way in which many of us initially think about capital: as monetary value. Social capital is the value of our human relationships and networks. Cultural capital encompasses modes and patterns of consumption and expression. Under certain conditions, social and cultural capital can be derived from economic capital through systems of value exchange. Context matters, though, as evidenced in Karun's case, for capital is also sometimes irreparably lost in conversion and translation.

Having capital makes certain opportunities in life more possible, or what we might refer to as privilege. Conceptually, privilege describes advantages only available to certain individuals and groups. One need not have earned the power that flows from privilege, or even be aware of it, in order to accumulate privilege over time and benefit from it.[29] Social class plays an important role in understanding privilege. Class both structures privilege and is a process through which privilege is produced as well as maintained. The structural view of class privilege defines social class through labels and hierarchical levels (e.g., working class, upper-middle class, or underclass), whereas the processual view defines it as an identity constantly shaped and reshaped by individual interpretations and shared experiences.[30] The descriptive categories under which privilege and inequality operate are themselves fluid and in perpetual motion. Distinctions between individuals and groups are both subjective and objective, with distinction being the capital that certain differences generate.

Bourdieu's conceptualization of cultural capital includes three forms as well: embodied, objectified, and institutionalized. Embodied cultural capital concerns learned ways of using one's mind and body, like the dialect or accent a person uses to speak. Objectified cultural capital involves the display of items and goods denoting status, such as a large collection of technological gadgets and obtaining the latest upgrades. Lastly, institutionalized cultural capital has to do with markers of official recognition and legitimation. This includes holding an advanced degree or set of credentials, and the use of any specialized terminology that only a degree holder might use. Bourdieu applied the theory of cultural capital to a range of "fields," or domains of life such as religion and law, but primarily focused on schools, arguing that institution plays the most significant role in reinforcing class relations.

Bourdieu's theorization of cultural capital is grounded in French schooling, status hierarchies, and signals of "high culture." It does not neatly map onto other cultural contexts and systems of legitimation.[31] Returning to Karun, she was a more privileged parent than most in Syria, intensely involved in her child's education. In the United States, both Karun and her husband were unemployed, and she was only as involved in Pargev and Joseph's learning as she could be considering her constraints. Schooling in the United States is deeply tied to middle-class cultural values such as independence and individual potential, and is designed to prepare children to participate in middle-class life.[32] In turn, educational reformers since the mid-twentieth century have blamed seemingly "uninvolved" working-class parents for declining schools. Cultural capital is widely used in the United States as grounds for social exclusion.[33]

Annette Lareau reoriented Bourdieu's class culture perspective to the US public education system, demonstrating how the class-based ideology at its foundation impacts working- and middle-class families differently.[34] Lareau draws a direct connection between class background and parental involvement in schooling.[35] She finds that social class shapes the cultural resources that parents have at their disposal to mobilize and influence their child's formal education. Beyond income, these resources include a network of college-educated individuals and professional work relationships. Child care centers, for example, introduce parents (and particularly mothers) to networks of opportunity. These benefits pay unexpected dividends in promoting child well-being for families across the socioeconomic spectrum.[36] Middle-class families have an easier time reliably activating these cultural resources, though, which enables them to build stronger connections between family and school. Challenging the misperception of uninvolved parents as unloving, Lareau finds that both working- and middle-class parents want their children to be successful.

Later research by Lareau has focused on variations in parenting styles.[37] She uncovered differing cultural ideas about child-rearing between working- and middle-class US families. These orientations are forms of what Bourdieu refers to as "habitus," or naturalized and internalized systems for structuring life. Middle-class families tend to follow a logic of "concerted cultivation," in that they value extracurricular activities and at-home learning experiences that nurture children's talents and interests. In addition to spending more energy meeting their children's basic needs, working-class families often orient their child-rearing practices around opportunities for "natural growth" such as unstructured play and time with neighbors. Public schools intentionally privilege middle-class approaches to parenting, and give middle-class children a "home advantage" at school. Lareau makes visible structural inequality in US public schools, and the complex dynamic between home and school life.

This work, however, centers primarily on mainstream classrooms. Sociocultural factors influence power imbalances between parents of students with disabilities and school personnel.[38] Audrey Trainor writes that "because participation in special education requires specialized types of cultural and social capital and occurs in a field with unique rules of engagement (i.e., habitus) meaningful participation is challenging to establish."[39] The US Department of Education reports that as of 2013, 13 percent of children ages three to twenty-one in primary and secondary US public schools (approximately 6.4 million) were receiving special education programs; of these children, 21 percent had speech or language impairments.[40] Upper- and middle-class parents have an easier time speaking the very complex language of special education, which includes knowing the latest therapies and how to prepare for important Individualized Education Program (IEP) meetings with their child's teachers, therapists, and school administrators.

In the years following Lareau's research, Internet penetration into children's homes has increased via broadband and wireless connections, and home has become the primary site of disabled and nondisabled children's increasing time spent with new media.[41] The iPads that children use as communication aids, and also as learning tools and fun toys, are but one technology among a constellation of other media that children and families use together. Parents borrow, purchase, and lease technology (e.g., books, computers, and Internet access) as a type of capital "investment" in their child's learning and down payment on future educational benefit.[42] These decisions are influenced by parents' cultural values, personal goals, and perceptions of their child's maturity in handling the responsibility of technology.[43] Social class alone cannot fully explain patterns of family life, but it can serve as a lens through which we understand the resources, strategies, and ideologies that give shape to family media practices, meanings parents and children associate with personal communication technologies, and ideas about the proper role of new media in society.[44]

Bourdieu wrote about "technical capital" as a subset of cultural capital (referring to manual workers' skilled use of machinery), but he did not discuss the networked and distributed skills needed to use information and communication technologies to one's advantage for improved opportunities in contemporary society.[45] Concurrently, technology can be thought of as a "strategic research site" for studying society and the organization of social practice.[46] It is not only technology but also the *culture of* technology that can reproduce social inequality.[47] In order to study how parents navigate their disabled children's iPad use, the technology must be understood within the system of social relations that shape and reshape its intended uses and cultural meanings.

Throughout this book, I empirically trace capital through technical and social domains, and make a theoretical contribution by linking understandings of capital and distinction across disparate work in education, disability, and technology. I concentrate in particular on the role of embodied, objectified, and institutionalized cultural capital in shaping how parents navigate their disabled children's use of mobile media and technology at home as well as the symbolic and material ways in which this use is tied to school and other institutions, such as health insurance companies and technology multinationals. In order to account for students with disabilities like Pargev in the US education system and society writ large, the "home technology divide" must also be inclusive of assistive technologies and assistive uses of off-the-shelf computers, as described below.

Reconsidering Assistive Technology

Not only do iPads subsume both pianos and computers; they are also technologies known as augmentative and alternative communication (or AAC) devices. Many nonspeaking or minimally speaking individuals such as Pargev use AAC devices to *augment* other forms of communication they might already use (e.g., nonverbal gestures and sounds such as laughter) and serve as an *alternative* to oral speech. AAC covers a diverse range of manual practices (e.g., American Sign Language) and variety of materials. AAC tools range from low-tech (e.g., plastic communication boards) to mid-tech (e.g., electronics with disposable batteries) to high-tech versions that allow individuals to convert text into synthetic speech (e.g., the computer used by physicist Stephen Hawking). Just as all of us triangulate our modes of interpersonal communication beyond oral speech—for instance, waving to friend or sending them a text with the "waving hand sign" emoji besides speaking the word "hello"—high-tech AAC is often used in combination with the other forms of AAC.[48]

It is difficult to get exact statistics on how many people communicate primarily through iPads and AAC apps. To provide a sense of scale, in one large school district in the Los Angeles area (with a K–12 enrollment of over 650,000), the district's lead AAC coordinator relayed that there were "at least 150" students using "iPads with apps" for AAC. The American Speech-Language-Hearing Association estimates that at least 2 million people in the United States have an impairment—whether from birth, or acquired later in life through an injury, illness, or progressive condition—that limits their ability to talk in the traditional sense.[49] From 2004 to 2014, the total number of children receiving public benefits for speech and language impairment increased 171 percent (from 78,827 to 213,688 children).[50]

AAC devices are traditionally categorized in the health and rehabilitation fields as a type of "assistive technology." The US Assistive Technology Act defines assistive technology as "any item, piece of equipment, or product system, whether acquired commercially, modified, or customized, that is used to increase, maintain, or improve functional capabilities of individuals with disabilities."[51] This definition encompasses an array of tools, from complex systems for accessing a personal computer through eye gaze input, to simple devices such as a magnifying glass for reading fine print. The US Individuals with Disabilities Education Act requires school districts to provide assistive technology to students with disabilities when it supports their acquisition of a free and appropriate public education, which is how Pargev ended up with a second, school-owned iPad with Proloquo2Go.[52]

In practice, though, the definition of assistive technology is vague. The International Classification of Functioning, Disability, and Health states "that any product or technology can be assistive."[53] This begs the question as to whether or not assistive technologies are exclusively for individuals with disabilities. Cultural theorists have understood artificial objects that mediate human subjectivity as "prosthetics," "technologies of the self," or things that enable the emergence of the human–machine "cyborg."[54] With little exception, these theorists rarely interrogate the lived experiences of disability as grounds for theory building.[55] Instead, Katherine Ott challenges scholars of technology and society to "[keep] prosthesis attached to people," and not ignore both the pains and pleasures that technology begets for those with disabilities.[56] Wheelchairs can provide comfort and ease of mobility, for instance, but prolonged sitting in one causes pressure sores.

The categorization of particular communication technologies as assistive and others as not is an inherently political choice.[57] While Apple's voice-activated interactive assistant Siri might be an assistive technology when used by people with disabilities, she is otherwise considered a more or less helpful personal assistant when utilized by able-bodied individuals.[58] This relationship between *assistive* and *assistance* automatically varies for each of us over our life span due to human growth and bodily degeneration. We fluctuate between degrees of independence from and dependence on other technologies (such as canes) and services (such as personal home care aides or our relatives). In fact, some cultural anthropologists argue that all human communication is in some way aided by assistive technology in the form of conversation partners and socially learned techniques that none of us are born knowing.[59]

Distinctions between mainstream and assistive technologies have material as well as symbolic consequences for people with disabilities. While their needs influenced the design of mass-market consumer goods and electronics in the late twentieth century through the philosophy of "universal design," individuals with disabilities tend to remain outside industrial designers' and engineers' imagined user base and the public-facing image of these products' promotion.[60] Assistive technologies have also historically been difficult for consumers to obtain and learn about because such knowledge tends to belong to specialized professional groups.[61] They are culturally associated with dependency and victimhood (i.e., the phrase "confined to a wheelchair"), which can negatively impact the way in which people with disabilities see themselves as technology users and how others perceive them.[62]

The immediate environment in which technology use is embedded, cultural factors impacting technology adoption, and dynamic qualities of both the technology and user all contribute to the social shaping of assistive technology.[63] With their exponential rise in ubiquity over the past two decades, mobile communication devices, as the next section details, are a significant site where the meanings of "mainstream" and "medical" technologies are being renegotiated.

Convergence of Mobile Media and AAC Devices

As a communication scholar, I was initially drawn to AAC devices because they are, by definition, mobile communication technologies (although they are not networked unless connected with Wi-Fi or cellular data). Clear plastic communication boards (also known as eye transfer boards or "e-tran") with the letters of the alphabet visible from both sides of the board are portable tools for creating shared meanings between a nonspeaking individual who spells words through eye gaze and their conversation partner who holds the board.[64] High-tech AAC devices in particular, though, provide a unique lens for reflecting on the emerging complexities of mobile communication technologies as well as their political economy.

In reference to the exciting potential around iPads as AAC devices, Mark, the father of River (age seven), observed, "I can't remember the guy's name, but one of the very first TED conferences, he introduced the touch screen and it just seemed like it would be used for more than just cash registers." The first-generation iPad debuted on April 3, 2010, a few months before I began my PhD program. In the days following the iPad's release, parent-uploaded YouTube videos of toddlers navigating the tablet's touch screen interface started to emerge online. This combination of nascent technology and nascent humans was a potent, user-generated marketing vehicle for the iPad. Noted one journalist of the newborn and new technology trend, "The litmus test for 'user friendly' until recently was 'Can my mom use it?' Increasingly it might become 'Can my toddler use it?'"[65] Like emerging media from the telephone to the television set before it, the iPad was linked from its inception to notions of conception and innovation.[66]

Mobile media are an increasingly integral part of many families' everyday lives in developed nations. The growth in the 1960s of domestic mobile communication technologies, such as portable telephones and televisions, reflected a transformation from what Raymond Williams once termed "mobile privatization" to what Lynn Spigel calls "privatized mobility."[67] While postwar telecommunications promised US suburban homes connections to the outside world, portable devices marketed to families decades later allowed home to follow them wherever they went. Today's mobile technologies, including iPads, both shape communication patterns and are integrated into existing ones for older media—a process referred to as "domestication."[68] There is no shortage of present-day ambivalence about mobile connectivity and family life, perhaps best illustrated by the 2011 book *Goodnight iPad* (a parody of the children's literature classic *Goodnight Moon*), which encourages children and their parents to power down as the sun sets.

Tablet devices running Apple, Android, or Windows operating systems can now be equipped with apps that mimic the software on "dedicated" AAC devices—dedicated in that their primary purpose is to aid oral speech. From a clinical standpoint, there are pros and cons to both dedicated and nondedicated AAC devices. Dedicated ones offer richer and more complex language software, but tablets are much lighter in weight. Many tablets have built-in cameras; one AAC specialist I spoke with called this feature a "game changer" because it allows users to customize the images that accompany vocabulary words within an AAC app (e.g., pairing a classmate's face with their name in the visual system), as opposed to taking a photo with a separate digital camera and uploading it to the dedicated AAC device via memory card or USB cord. Assistive technology companies such as Dynavox and Prentke Romich that produce dedicated devices have robust customer service divisions, but repairs to broken devices can take months; a busted iPad can be replaced with a quick trip to the nearest Apple Store.[69]

While dedicated AAC devices have traditionally cost thousands of dollars, less expensive and commercially available tablets have unsettled the AAC market.[70] At a price tag of $250 in 2016, however, Proloquo2Go is still one of the most expensive apps in the App Store—almost as costly as the least expensive new iPad, a 16GB Mini 2 going for $269. AssistiveWare does not offer a free trial version of the app. So while the combination of the iPad and Proloquo2Go is less expensive than a dedicated device, even the app itself is cost prohibitive for many families, especially considering that the general out-of-pocket costs incurred by a family raising a child with a disability are already quite substantial.[71]

AssistiveWare was founded in Amsterdam in 2000, and has been intimately linked with Apple from its inception. The company exclusively develops products for Apple's mobile and desktop operating systems. It initially released Proloquo2Go in April 2009, prior to the debut of the iPad, at first for the iPhone and iPod Touch. In 2009, Apple named Proloquo2Go one of the top thirty apps of the year in its iTunes Rewind.[72] AssistiveWare made Proloquo2Go 1.3 available for the iPad in April 2010, shortly after the hardware's release, as a free upgrade for users of the app on iPhone and iPod Touch.[73] In May 2015, the app became integrated with the wearable Apple Watch for use across multiple mobile devices.

Youth with communication disabilities and their families represent a growing market for apps and tablet-based AAC devices in general, and iPads in particular. A 2014 market survey found that the Apple iPad was the number one brand among US children ages six to twelve, topping all other consumer products.[74] Just between 2011 and 2013, tablet computer ownership among families with children ages eight and under increased dramatically from 8 to 40 percent.[75] While overall ownership of tablets is on the rise among families, there are substantial divides by income. Among families with a combined household income of $100,000 a year or more, two-thirds (65 percent) own a tablet computer, while among families earning less than $25,000 a year, ownership is only at 19 percent.[76]

Apple has a storied relationship with parents of disabled youth. In the early 1980s, Apple was one of the first computer companies to have an internal group dedicated to accessibility. In 1986, it partnered with the Disabled Children's Computer Group, a Bay Area organization comprised of well-resourced, tech-savvy parents of youth with disabilities advocating for their children's needs as computer users.[77] Apple, however, also had paternalistic motives in forging this alliance. The company did not target individuals with disabilities as a wider market but instead as beneficiaries of the company's charity and goodwill.

Apple has taken a similar approach to its association with Proloquo2Go and parents of children who use the app. A 2013 Apple marketing campaign featuring a Proloquo2Go user—Enrique Mendez, mentioned at the start of this chapter—and testimonials from his family claims that the company is "Making a Difference. One App at a Time"—that difference being an undoubtedly positive and constructive one.[78] Through the partnership, AssistiveWare receives major publicity and Apple gets to portray its brand in a flattering light. While Apple bills itself as representative of creativity, communication, and freedom, the company also heavily constrains what users can do to alter their hardware and software.[79]

The short film also conveniently omits three important details, explained below: the time and labor-intensive process by which parents of children with communication disabilities attempt to obtain iPads as AAC devices, market-driven political economy of schools' selection of educational technology, and complexity of Proloquo2Go as a communication technology.[80]

Traditionally, a child in the United States receives an AAC device (be it an iPad with Proloquo2Go or a different system) only after licensed specialists conduct clinical assessments, manage periods in which the device or multiple different devices are used on a trial basis, and write recommendations to schools (if the AAC system is deemed educationally necessary) and insurance agencies (if medically necessary) to ultimately fund an AAC device. In the United States, families usually play a significant role in this selection process, but there are significant barriers to participation.[81] These include culturally and linguistically inaccessible forms of parent training, biases in funding processes, and technical difficulties in learning to operate the hardware and software.[82]

Seeking financial support for obtaining a speech-generating device can also be an overwhelming, frustrating, and challenging process. The cost of AAC devices generally exceeds a user's ability to pay for it on their own. Each public and private funding agency sets its own terms for eligibility, requires particular formats and wording in their documentation (e.g., a letter of medical necessity written by a therapist or state-specific certificate of medical necessity), and demands multiple steps in application processes, all of which inevitably leads to missteps and request denials.

US government provisions for therapeutic services and assistive technologies also differ greatly based on personal characteristics such as age (e.g., birth to three) and disability status (e.g., multiple disabilities) as well as by state and school district. Unlike adults, who do not have blanket entitlement to speech-language supports, school-age children in the United States are able at least in theory to claim educational necessity for their AAC devices under the Individuals with Disabilities Education Act when their communication limitations are so significant that they impact children's access to, participation in, and potential to demonstrate progress in the general curriculum, extracurricular activities, and other nonacademic activities.[83]

Yet insurers have been resistant to fund tablet-based AAC devices. While Medicare considers speech-generating devices to be "durable medical equipment," this does not extend to personal computers used as AAC devices, only the speech-generating software that individuals download onto their computers.[84] Government and private insurers fear the fragility of the iPad when used for constant communication, liability risks, and potential for fraud and resale.[85] Medicare's stated explanation is that tablets "are useful in the absence of an illness or injury"—as are popular wearable health technologies such as fitness trackers produced by Fitbit—and thus cannot be classified as durable medical equipment.[86]

Medicare coverage also stipulates that the technology must be "limited to use by a patient with a severe speech impairment."[87] This reflects a cultural value of individualism that may not mirror the ways in which new forms of AAC devices are being used collectively in families. In a 2015 survey, 71 percent of families that own a mobile media device with an AAC app installed report that other family members had access to or used the device besides the person using AAC to communicate.[88] Along with apps for AAC, tablet-based devices provide a wide range of other apps for social media, communication, and entertainment.[89] A 2012 survey conducted by AssistiveWare indicated that 90 percent of people using iPads and iPods for AAC used the device for purposes besides speech output as well, such as Twitter, Facebook, and YouTube.[90]

In the state of California, possible funding sources for a school-age child's AAC device includes California Children Services, MediCal, private health insurance, general public school budgets for funding assistive technology, and philanthropic organizations. Children with a low-incidence disability (e.g., cerebral palsy) in California can have their iPads with Proloquo2Go funded by the school district through a special assistive technology fund, while children with a more frequently occurring disability (e.g., autism) tend to have their devices paid for by the child's school. Though sales figures for Proloquo2Go are not publicly available, its impressive iTunes ranking (number forty on the list of top-grossing iPhone apps in the education category as of June 2016) is explained in part by bulk education sales. School districts get a sizable 50 percent discount on Proloquo2Go by purchasing twenty or more licenses of the app through Apple's Volume Purchase Program for educational institutions.[91] Districts nationally are uneven, though, in offering iPads for AAC as well as training speech-language pathologists, teachers, and staff on how to use them.[92]

It is important to note that Proloquo2Go requires technical expertise, digital literacies, domain knowledge, and comprehension skills to understand and use. This explains the need for speech-language pathologists with additional education in assistive technology like Rachel and Caren to provide one-on-one at-home training to parents in how to support their children's use of the app. While explaining the app's full functionality, interface, and design is far beyond the scope of the book, figure 1.1 illustrates the main screen, or Grid View, of Proloquo2Go. It is one of three interface options for speaking with the app, including Recents View (which provides a shortcut to a set of messages recently spoken) and Typing View (which displays a text pad for speaking through words manually typed by the user). In Grid View, a home page contains a mix of buttons and folders.

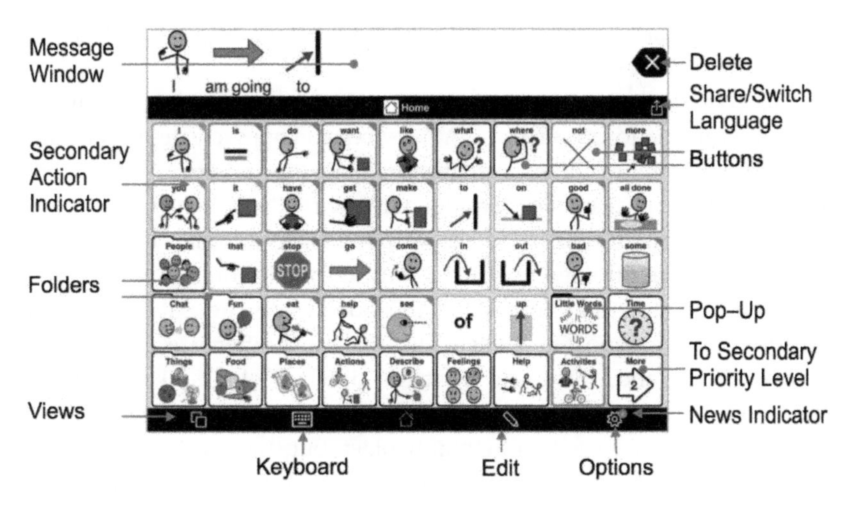

Figure 1.1
Proloquo2Go's Grid View, with explanatory labels as indicated in the Proloquo2Go user's manual. Image copyright AssistiveWare B.V. Used with permission.

Folders contain buttons grouped by categories, such as "Feelings" and "Places." The Message Window serves as a sort of drafting board for users to string words and phrases together with buttons. Once the user is done drafting their phrase or sentence, they can press the white space of the Message Window for the message to be spoken through synthetic speech output.

At the beginning of my fieldwork, I was perpetually confused by Rachel and Caren's use of the phrase "programming the device" in reference to Proloquo2Go and the iPad. I wrongly assumed that it had something to do with coding in a computer programming language. "Programming the device" did involve computers and language, though; the term referred to the continual process of maintaining and making changes to which preloaded vocabulary was included in the AAC system (including adding, deleting, and modifying existing vocabulary words) as well as determining how best the vocabulary should be visually organized so that the AAC user could easily navigate the system independently.[93] Although Proloquo2Go comes with three preset vocabulary configurations ("Basic Communication," "Intermediate Core," and "Advanced Core"), these setups are endlessly customizable. One person's set and arrangement of vocabulary rarely exactly matches that of another user. The potential for disorganization and duplicate vocabulary entries in the system increases with more individuals going into the app and making programming changes, including parents, therapists, and teachers that regularly communicate with the child.

The technological, economic, and cultural convergence of mobile media with other media forms, such as AAC devices and apps, creates opportunities for some families and exacerbates challenges for others.[94] When privileged parents buy iPads and Proloquo2Go with their own money, they circumvent public funding and school purchasing schemes. Considering the opportunities afforded by innovations in mobile communication and yet significant structural limitations, this book traces the extent to which the iPad and Proloquo2Go are actually "making a difference" in families' lives, for better and worse, and what this difference looks like across the socioeconomic spectrum. Parents and children may learn to navigate Proloquo2Go's Grid View with a helpful booklet, but there is no user manual for traversing the complex political and cultural conditions of raising young people with disabilities, or growing up with a disability, as the next section details.

Parenting Digital Youth with Disabilities

The contemporary role of media and technology in the lives of children with communication disabilities and their families must be understood within the context of particular US policies as well as historical conditions surrounding disability and parenting. Prior to the 1970s, US law actively suppressed disability in public spaces through the enforcement of "ugly laws" that barred "unsightly beggars" from city streets, eugenics laws that led to the institutionalization and forced sterilization of disabled adults and children, and laws prohibiting children with disabilities from entering public schools.[95] Psychologists thought parents caused their child's disability, and promoted the removal of children from their families as cures.[96] By 2010, though, only 4 percent of those living in residential settings were age twenty-one and younger, compared with 36 percent in 1977—a shift accelerated by the passage of the Olmstead Act of 1999, which stated that the unjustified segregation of people with disabilities violated the Americans with Disabilities Act of 1990.[97]

Discussions about parenting a child with a disability are also inexorably gendered.[98] Over the past decade, feminist disability studies scholarship has invited reexamination of the meaning of motherhood over history, and the ways in which media narratives reflect and shape the lives of families of children with disabilities.[99] One infamous example is the "refrigerator mother theory," the largely discredited yet persistent Freudian-inspired concept developed by child psychiatrist Leo Kanner and popularized by psychoanalyst Bruno Bettelheim.[100] The theory alleges that "cold" mothering and women's career aspirations outside the home lead to childhood autism, and posits a causal link between the influx of domestic technologies (such as refrigerators) in the postwar US home and a perceived societal devaluing of mother–child relationships.[101]

While Kanner and Bettelheim's claims are widely considered suspect, mothers in the twenty-first century are still blamed in other ways for their child's disability.[102] For instance, mothers of autistic children are admonished for having vaccinated their child, not being vigilant enough in noticing early signs of their child's autism, and insufficiently seeking out and administering the latest therapies and treatments.[103] The United States is in the midst of a cultural shift away from the refrigerator mother archetype toward an "intensive mothering" paradigm.[104] Amy Sousa writes that "warrior-hero mothers are now responsible for curing the disability, or at least accessing the intervention that will mitigate the disability's impact on their children."[105]

Both the refrigerator mother and warrior-hero mother scenarios, however, define disability as something to be eradicated, and view children with disabilities as burdens to their parents. One alternative to the language of tragedy can be seen through resiliency theory in the field of social work. Resiliency theory puts forward the idea that families of children with disabilities generally develop accommodations, or "proactive efforts of a family to adapt, exploit, counterbalance, and react to the many competing and sometimes contradictory forces in their lives."[106] This might include preparing separate meals for the child with a disability, or making sure the doors of the home are always locked if the child has a tendency to wander.

Another type of accommodation that families of children with disabilities make is altering their media and technology use.[107] Some accommodate for behavioral difficulties on car rides by providing backseat DVD players. Others make changes in their home television viewing habits, including having separate screens for different family members, watching child-oriented programming together, or not watching television at all.[108] Due to sensory, hormonal, and neurological issues, some children have difficulty sleeping; children who cannot fall back to sleep may turn to media for comfort. For many of these families, and those with disabilities themselves, disability can be a source of pride as well as a positive aspect of individual and collective identity.[109]

Digital Media and Disability

The approach to parenting, media, and technology in this book bears the influence of critical disability studies as well as work from media, communication, and science and technology studies that incorporate disability perspectives.[110] Disability studies scholarship in the United States is rooted in civil rights activism of the late twentieth century, following in the tradition of critical race, gender, and sexuality studies while building on this work as well.[111] Much disability studies scholarship pushes back against the "medical model" of disability, which is grounded in the assumption that disability is an individual biological burden or deficit.[112] In response to the medical model, some disability activists and scholars emphasize a "social model" of disability. This model holds society accountable for shaping the lived experience of disability and its potential to enhance or detract from an individual's life as well as our collective culture.[113] The social model generally makes distinctions between impairment (bodily difference) and disability (the social and built environment that disables different bodies).

A number of scholars drawing on feminist and poststructuralist theory critique the social model, though, for drawing clear differences between impairment and disability, akin to a false sex/gender binary.[114] Rosemarie Garland-Thomson notes that all bodies, depending on the environment, situation, and interaction, have "varying degrees of disability or able-bodiedness, or extra-ordinariness."[115] Alison Kafer further complicates this view with work from the environmental justice movement, writing that "disability is more fundamental, more inevitable, for some than others: the work that one does and the places one lives have a huge impact on whether one becomes disabled sooner or later, as do one's race and class position." Drawing on queer and feminist theory, Kafer instead offers a "political/relational model of disability," in which disability is a set of political practices and social associations—"a site of questions rather than firm definitions."[116] Feminist disability theory stresses that disability is experienced in and through relationships, is bound up with the lives of people with and without disabilities, and that fighting back against discrimination requires coalition building.

In the digital age, a variety of technological, political, and economic barriers limit the agency of individuals with disabilities and their families.[117] Various platforms, applications, and websites strongly discourage individuals with disabilities from cultural as well as societal participation. This includes files and websites that cannot be properly read by the screen readers of blind and visually impaired individuals, and YouTube videos sans captioning or with poor auto captioning viewed by Deaf audiences.[118] Inaccessible technology helps very little and, in fact, creates new forms of exclusion where none existed before.

At their core, these incomplete remedies are based on a seductive belief in the easy technological fix as well as a view of individuals with disabilities as most in need of fixing—whereby technology repairs or eliminates impairment.[119] In a technologically determinist version of this relationship, technology alone enables disabled individuals to overcome disability and serve as inspirations for nondisabled people; in a socially determinist form, visionary technologists liberate individuals with disabilities from the constraints of their minds and bodies. These narratives take on new meanings among parents, teachers, and therapists. As one AAC specialist I spoke with noted, "Sometimes parents hold on to. ... They want a thing. They want a device to help, to fix their child. Which, it's not. It's a tool, and all tools are human dependent." Dismantling determinist views of the relationship between technology and disability requires examining up close the range of mediated encounters had by disabled individuals across the socioeconomic spectrum, including youth with disabilities and their families. This reflects Williams's call to reject both "technological determinism" and "determined technology."[120]

Empirical research on the well-documented social and cultural "participation gap" among youth has been quite limited with respect to disability.[121] Outside classroom and therapy settings, we know very little about the experiences that disabled youth, their siblings, and their parents have with media and technology at home and as part of domestic activities.[122] While research on how class, gender, sexuality, and race shape new media use among young people is growing, Gerard Goggin writes, "There has been even less work on disability, youth, and mobiles, with the research literature focusing still on issues of accessible design, or hamstrung by outmoded accounts of impairment and disability."[123] This book takes up Goggin's call, identifying intersecting issues of privilege and oppression that affect the lives of youth with disabilities in their engagement with media and technology.

Intersectionality and Distinguishing Parents

The caregivers of youth with disabilities tend to be grouped together under the umbrella of "special needs parents." In her ethnographic work with US mothers of children with disabilities, Gail Landsman saw no significant difference by socioeconomic class or education level in terms of how these women constructed their identities. She found that mothers instead believed they were "in a class by themselves" compared to mothers of non-disabled children, due to their distinct child-rearing experiences. Moreover, Landsman suggested that "discrimination against persons with disabilities extends broadly across class lines in U.S. culture."[124]

Children with disabilities and their families also represent this country's racial, ethnic, socioeconomic, cultural, and linguistic diversity, and as such, may have relatively little in common with one another. Over the past decade, parent-reported childhood disability in the United States has steadily increased to nearly six million children under age seventeen, growing by 15.6 percent between 2001–2002 and 2010–2011.[125] Family incomes below the US federal poverty level are associated with a higher prevalence of parent-reported developmental disabilities.[126] Under the specter of extreme income inequality in the United States, policies impacting health insurance, housing, food insecurity, minimum wage, and costs of child care profoundly impact these families. For instance, autistic children and those with other developmental disabilities from immigrant families were more than twice as likely as nonimmigrant families as of 2012 to lack consistent health care, and three times as likely to lack any type of health coverage at all.[127]

Studying the experience of disability in the digital age and especially among families requires an intersectional approach.[128] Intersectionality, as a concept, emerged in the late 1980s and early 1990s as a way to critique academic work that focused either on race or gender in isolation, and pushed black women away from centers of power.[129] While a sweeping discussion of the benefits and drawbacks to intersectional analyses along with its grounding in work by feminists of color is beyond the scope here, it is important to acknowledge the far-reaching applications of intersectionality theory to the study of intergroup and intragroup relations.[130] The advantages and disadvantages of different types of privilege are not simply additive or subtractive.[131]

Patricia Collins instead emphasizes the significance of dynamic centering and relational thinking, which consists of placing two or more systems of power at the middle of an analysis, and asking how they shape one another.[132] For my purposes here, I focus more squarely on the dual distinctions of disability and class, while also attending to how individual identities and institutional factors interact with age, gender, race, ethnicity, immigration status, and linguistic background. For example, the cultures that immigrant parents like Karun come from frame how they view their children and understand disability. I forefront the simultaneity of these identities and their fluidity, with parental privilege relative to the context of the greater Los Angeles area.

With that, I identified four characteristics of more and less privileged parents in my study, centering on mothers' education level, household income, English-language fluency, and ownership of the iPad that the child used for AAC. Table 1.1 provides a list of the names of children and parents observed or interviewed in each group.[133] Of the twenty families I studied, ten were more privileged and ten were less so (for further discussion, see the chapter on methods).

First, in more privileged families, mothers tended to be college educated, while those in less privileged families had more often completed high school at most or had taken some college classes. Second, more privileged families tended to have a combined yearly household income of $100,000 or more, whereas less privileged families generally indicated earning $50,000 per year or less—below the median income in all three counties under study in the Los Angeles area.[134] This is not to say that income is a fixed variable. Some parents, like Karun, experienced fluctuations in their economic stability due to factors such as divorce or illness. Nelson explained that before the 2008 economic recession in the United States, "I used to make more money," but with his wife's cancer and his daughter's autism diagnosis, the job could not accommodate his family's needs. He found a

Table 1.1
Members of Less and More Privileged Families

Less privileged families		More privileged families	
Child	**Parent/s**	**Child**	**Parent/s**
Paul Michael	Garine and Levon	Nash	Taylor and Todd
Beatriz	Pilar and David	Thomas	Daisy
James	Cathy	Raul	Nina
Madeline	Teresa	Luke	Debra and Rob
Stephanie	Marisa and Nelson	Danny	Alice and Peter
Pargev	Karun	Isaac	Sara
Talen	Kameelah	Eric	Anne
Kevin	Rebecca	Chike	Esosa
River	Mark*	Cory	Perri
Moira	Vanessa*	Sam	Donna

*Single parent

job that while paying less, allowed for more flexible hours. "I'm only here," he explained, "because this job gives me the freedom to be with my family whenever I have to be."

Third, both more and less privileged parents tended to be fluent in English, although this commonality is exclusively due to my own sampling bias. One requirement for participation in the study as well as a limitation was that at least one parent needed to be fluent in English. This was owed to my lack of non-English-language skills and limited funds for a translator during my dissertation research. In a couple of families (Beatriz and Stephanie's), though, one parent (all fathers) translated for the other one.

Fourth, greater numbers of more privileged families (six out of ten) owned the iPad that their child used with Proloquo2Go (as opposed to it being school owned) than did less privileged families (two out of ten). Of those two families, one had saved up money to buy the least expensive iPad Mini for their child (River), and the other had received the iPad and Proloquo2Go from a charity (Pargev). As more privileged parents had greater amounts of discretionary income, they also tended to have other Apple devices at home onto which Proloquo2Go could be installed as a backup. The AAC coordinator for one district explained, "Sometimes I have families that will have an iPad with Proloquo2Go loaded at home, and we have the school one. They don't want the school one to be sent home. What we will do is create a common Dropbox account so that they'll have the same vocabulary on both iPads."

More privileged parents also tended to view iPads as easily replaceable should the technology break. For instance, Peter, father of Danny (age six), remarked, "Quite frankly, if Danny destroys this tomorrow, I can go buy an iPad 2 for 400 bucks that runs." Rob and Debra mentioned how a few months ago, their house had been broken into and someone stole the iPad that their thirteen-year-old son, Luke, was using on loan from school. Rob said, "I suppose, technically, [the school] would have to supply it," but instead he went out and immediately bought another iPad. "There was no way I was going to wait for [the school]," explained Rob. "I had to quick order one, get it down here, and get it, that kind of thing."

In contrast, less privileged parents were more often fearful of something happening to the iPad, and being on the hook for replacing or repairing the broken technology out of their own pockets, per their loan agreement with schools and regional centers.[135] One assistive technology specialist in a predominantly low-income and Hispanic-Latino neighborhood in Los Angeles mentioned that the parents they worked with tended to hesitate about bringing home the iPad that their child used to communicate in school:

"We said, 'Don't worry, please take it, use it.' And they said, 'Oh, OK, because we don't want anything to happen.'" That specialist's apprentice concurred, noting, "We've had parents nervous that their kid would be taking it on the bus to go home, and they're worried that another kid would steal it on the way home. They wanted to put a lock on their backpack." Their remarks as well as the others above make it clear that we cannot refer to parents of disabled youth as being in "a class by themselves" without unpacking the class distinctions and other forms of difference among them.

Giving Voice

As I mentioned at the start of this chapter, reports about the use of tablets and smartphones as speech-generating devices explicitly call on the phrase "giving voice" to highlight narratives of personal liberation via technology. Such accounts have been widespread in the North American news media since the launch of the Apple iPhone in 2007 and resulting market for AAC apps.[136] These news stories portray technology as allowing individuals to "overcome" their disability as an individual limitation, and are intended to be uplifting and inspirational for able-bodied audiences—both common themes in the mass media historically.[137] Consider Microsoft's Super Bowl ad in 2014, which features former NFL player Steve Gleason, who lost the ability to produce oral speech due to ALS. The commercial claims that the Microsoft Surface Pro tablet computer "has given voice to the voiceless," exemplified by Gleason providing the ad's voice-over, with the implication that he was once voiceless but now has voice thanks to Microsoft's innovations (figure 1.2).[138]

Mass as well as social media are implicated in the valuing and devaluing of voice. Spokespeople, celebrities, elected officials, and public figures serve as mouthpieces by speaking on behalf of others. The *Code of Ethics* of the Society of Professional Journalists explicitly states that a key journalistic duty is to "be vigilant and courageous about holding those with power accountable. Give voice to the voiceless."[139] Yet mass media give more lip service than voice by perpetuating the essentialist notion that being "voiceless" is a stable and natural category. Social media differ from the press by giving voice en masse, but this collective sounding board can be both productive and destructive when speech is used for peaceful or violent means, such as with online harassment and networked misogyny.[140]

A "sociology of voice" interrogates the structural conditions that strip humans of their humanity as well as their right to communicate.[141] This book offers a new angle on established critiques of how communication

Figure 1.2
Screen grab from a YouTube video of Microsoft's 2014 Super Bowl commercial. Former NFL player Steve Gleason, who lost the ability to produce oral speech due to ALS, narrates that Microsoft's technology "has given voice to the voiceless."

technologies give or limit voice through the case of individuals with communication disabilities who rely on mobile devices for speech. Jo Tacchi notes that the consequences of digital media for voice are not bestowed on the technology itself but instead enacted in contexts.[142] In turn, what if we *accounted for* and were *accountable to* those unable to produce, or who have significant difficulty producing, embodied oral speech, or what one might traditionally call "talking"? How useful is this figure of speech if we are to be fully inclusive of all citizens? What kind of discursive work does "giving voice" do? Does this converge or diverge from the meanings and practices that AAC users along with their conversation partners associate with the iPad—a tool that converts physical actions like button pressing into audible voice?

Mobile communication technologies can exacerbate rather than reduce inequalities. This is particularly true among "underconnected" children and their parents, who get by with intermittent mobile-only Internet access through one or a few smartphones, but face difficulty in accomplishing complex tasks for work and school that are better suited for a computer.[143] While mobile media are widely hailed as the most accessible tools to give voice to marginalized populations—due to their pervasiveness, low cost, and ease of use—I found that far from being equalizers or amplifiers, such tools unintentionally contributed to naturalized disempowered states and exclusionary positions, such as being "voiceless," "speechless," and "silent." The notion that mobile media "give voice" masks assumptions about ability, embodiment, and difference in the design, construction, and study of sociotechnical systems.

Though "voice" is problematic in its own right, it is also important to highlight how the terms "verbal" and "speech" tend to imply normative associations with the body and orality. It is common clinical practice to refer to users of speech-generating devices as "nonverbal." But that phrase does not reflect all the ways in which individuals with communication disabilities engage with the world of words through various media and technology. In a discussion of people who communicate primarily by typing on a keyboard, feminist scholar Lisa Cartwright explains, "Here we have an obvious double mediation: the computer and the human hand mediate speech in the place of the normative technology of speech, embodied oral voice."[144] I draw on Cartwright and employ the term "embodied oral speech" (i.e., people who have difficulty producing embodied oral speech) instead of "nonverbal."

I also focus deliberately on the experiences of parents in managing their non- and minimally speaking children's media and technology use. Kathy L. Look Howery writes that "unlike a child who has learned to speak naturally and therefore in a true sense 'have' their words, ... a child who uses [a speech-generating device] is given their words. Parents, therapists, or teachers put vocabulary (words) into the devices, when children are learning to use their [devices] they must find the vocabulary that others have given to them."[145] While parents do not put words into their nonspeaking children's mouths, they do have a unique relationship to the linguistic equipment that their children use to communicate. I embrace a child-centric approach that takes context seriously and does not overemphasize the individual. By concentrating primarily on parents, and secondarily on therapists and teachers, I complicate the idea that children's voices, both disabled and nondisabled, somehow exist in a state free from adult influence (including researchers), and push beyond a simple dichotomy of adulthood versus childhood.[146] The metaphor of voice more often reproduces than repairs imbalances of power and knowledge.

Overview of the Chapters

It is worth noting that this is the first book-length study of iPad use and adoption, not just among people with disabilities and their families but in general. Anyone interested in smartphones, tablets, and mobile communication will be interested in the chapters that follow due to the close attention paid to the unfolding developments of this technology, its influence across the business, health, and education sectors, and how individuals use, domesticate, negotiate, and shape their media tools. As opposed to lengthy industry-focused white papers on new technologies, which tend to paint broad strokes with a positive spin, I take an expressly critical and social scientific approach that situates the device in specific everyday contexts.

I found that parents' understandings of their child's iPad use aligned with or differed from popular, professional, and institutional definitions. Moreover, the function of economic, social, and cultural capital in parents' meaning-making processes varied. Using grounded theory, I identified five key areas (one for each main chapter) of difference in parents' understanding of their child's use of the iPad for AAC. Each chapter details how more and less privileged parents articulate what the iPad means to them, how these meanings shape management of their child's media and technology use, and how these conceptions both conflict with and complement dominant discourse about technology "giving voice."

Meaning making is an ongoing practice of turning *things* into *things that matter to people* through social and psychological transformation.[147] I make use of an expressly social definition of media, per Lisa Gitelman, who refers to media as "socially realized structures of communication," encompassing cultural forms, learned techniques and protocols, and shared practices.[148] In addition, I employ Don Norman's notion of the "conceptual model," or a mental simulation of a given piece of technology that designers and users each develop.[149] The conceptual model serves as a reference point for how users think media and communication technologies should be interacted with, or what they are designed to do. While this cognitive representation is in part due to a technology's physical properties, a person's conceptual models for a device such as an iPad also emerge from their life experiences, understandings of social norms, and cultural context.[150]

Each chapter thus poses a basic question about how parents interpreted one aspect of their child's iPad, framed at the beginning of each chapter with a provocative quote or two from parents directly. Chapter 2 explicitly poses the question, "What is voice?" As noted above, voice reoccurs as a rhetorical trope over history in popular discourse about nonspeaking individuals as well as AAC devices. This reduction of voice obscures the ways that privilege is built into sociotechnical systems—both programmed into the software itself and embedded into social practice. More and less privileged parents each constructed different meanings of voice in relation to the iPad, their child, and the perceived possibilities and limitations of the speaking world.

Chapter 3 asks, "What is a mobile communication device?" and responds with an unexpected answer. While it has been well documented by communication scholars that mobile devices have symbolic meanings beyond the messages they transport, I illustrate in this chapter how the protective case *around* a mobile device, an otherwise-unremarkable and forgettable object, is itself a key visual and material signifier of the sociotechnical world. I discuss how the choice of an iPad case became a major source of frustration and site of negotiation between school districts and parents. The iPad case reflected various tensions in how children with disabilities are perceived: between normalization and the child's "special" status as well as whether the goal of the case was to protect the computer or empower the child over the long run. Working-class youth were the most vulnerable to mercurial school district policies that valued their investment in the technology over a sustained buy-in toward students' futures.

In chapter 4, the question is, "What is an iPad for?" Regardless of their socioeconomic status, most families in my study somehow ended up with two iPads in their household. All families distinguished between the two in some manner. More and less privileged parents differed, however, in where they drew these boundaries—one iPad for "fun" and one for "communication" (among more privileged parents), and one iPad for "education" and one for "entertainment" (among less privileged parents). Their conceptual models and social understandings emerged and diverged partly due to how the iPad is designed and manufactured, but also due to class differences in the regulation of children's technology use in public and private spaces.

In chapter 5, I inquire, "What does it mean to communicate with an iPad?" Clinicians promoted Proloquo2Go as the "proper" way to communicate using an iPad; yet many parents interpreted their nonspeaking child's recreational use of iPad apps and other media as expressing socioemotional, cognitive, and verbal skills—complicating the fun/communication binary explored in the prior chapter. I look at the need to shift perceptions of both disability and children's popular culture away from having to do with deficit and deficiencies (i.e., disability as a lack of ability and kids' media culture as lacking educational value), and instead toward more asset-oriented models. It is nevertheless important to note that less privileged parents faced greater difficulty than more privileged parents in parlaying their children's expressive media use at home into recognition of value among teachers and therapists.

Chapter 6 examines the question, "How do media shape understandings of the iPad?" This chapter analyzes parents' interpretations of cultural representations of the iPad, Proloquo2Go, and AAC as well as their participation in the consumption, circulation, and creation of media about disability, parenting, and assistive technology. More and less privileged parents engaged in some similar but also strikingly different media practices. While the "disability media world" may be expanding, it also remains largely dominated by those parents with access to more distinctive forms of social, economic, and cultural capital.[151]

Lastly, in chapter 7, I summarize the ways in which more and less privileged parents' lives converged and diverged—not only in terms of their approaches to the iPad, but also in how they conceived of themselves and their children as representative or unrepresentative of the other families participating in the study. Researching kids' and families' media and technology use through the lens of intersectionality, I argue, allows for more effective coalition building among families pushed to the margins and otherwise silenced.

Conclusion

During a coffee break at a small, local, assistive technology conference in Southern California, I chatted with a more privileged parent named Donna, the mom of Sam (age seven), a nonspeaking boy with multiple disabilities, including autism and spina bifida. Sam had used Proloquo2Go on the iPad for about a year, but then switched to a type of dedicated high-tech AAC device known as a Vantage Lite. Donna told me how Sam used his Vantage Lite to speak to his grandmother remotely. "Each and every night," Donna shared, Sam "uses FaceTime through the iPhone. We place the phone through the handle [of the Vantage Lite] so it stands up, and he talks to his grandma and tells [her] about his day." The video chat does not bridge a long distance; in fact, Sam's grandma "lives five miles away," said Donna. "They converse and it's all for practice. ... It's a nightly ritual that we do."

Karun described how her extended family had participated in a similar ritual back in Syria. As part of Pargev' homeschooling, Karun taught him "communication by mobiles." She started out by having Pargev memorize phone numbers that she had written for him in a little address book. "We would put mommy's number, daddy's number, grandmother, grandfather's number, all the numbers," Karun recounted. After Pargev had memorized them, Karun wrote each phone number down on a little scroll of paper.

Each day, as part of his lesson, Pargev had to select one scroll and call the number.

Like Donna, Karun enlisted both her social network and the telecommunications network in order to distribute support for her son's learning. She tasked each family member with asking Pargev questions like "'How are you? What did you do? Did you do math? Where are you going today? What day is today? What time is it?' Things like that, just to make him get used to talk[ing] on the phone." Whereas Donna currently found it easy to implement daily chats between Sam and his grandmother, Karun spoke of their family's routine strictly in the past tense. "I used to do things like these things a lot over there," she told of life in Syria. Now, the family had to live within a much more constrained set of economic, social, and cultural resources. "For my husband, he left his own clinic. We left our parents. Everything else we left there," said Karun.

While smartphone owners increasingly use these tools for much more than telephone calls, their association with vocalization persists. A cadre of contemporary critics, most notably MIT professor Sherry Turkle, frequently pen articles and give interviews contending that (to use Karun's words) "communication by mobiles" is replacing interpersonal communication in everyday life.[152] They say that too often, technology is a communication replacement, not an enhancement. Clinical psychologist Catherine Steiner-Adair argues that our handheld devices get in the way of authentic connections and erode "the art of talking."[153] Turkle warns of phones at the dinner table being a distraction as well as diluting conversation even if they are turned off.[154]

This binary between face-to-face and mediated communication is patently false and further complicated by individuals who primarily "talk" using mobile media and employ communication technologies that both augment and provide alternatives to their oral speech production.[155] The mobile device on the kitchen table, in the form of an iPad propped up with the Proloquo2Go app open, does not degrade an empathic bond between family members so much as potentially enable it in the first place. How Sam, Pargev, and their families experience connection and disconnection on a societal as well as interpersonal level cannot be reduced to the technology, or their disabilities, alone.

In the chapters that follow, I argue that the sociocultural, political, and economic institutions within which families of children with disabilities are embedded shape the role of new media in their lives. I investigate how boundaries are maintained between the home and outside world, between public and private spaces, and between the iPad screen and screens of other technologies that families regularly use. Parents' class background influences how they understand the value and purpose of the iPad as well as their relationships with the social entities shaping the use and deployment of the technology at home. Tablet-based AAC devices have incredible potential to support agency, independence, and personhood, but they do not enter into a vacuum devoid of other injustices. They become part of a system reproducing structural inequalities; nonspeaking children whose parents are best able to leverage social, cultural, and economic capital to navigate the bureaucracy tend to benefit the most. Essentially, when individuals adopt and use communication technologies that are expected to "give voice," they frequently get much more than they bargained for.

2 Talking iPads and the Partial Promise of Voice: What Is Voice?

> The way I equate it is that you don't not give someone a wheelchair. You don't not give someone glasses. You don't not give someone crutches. It's the same as a voice. Your voice is an extension of your body. ... I don't understand how people can't understand that [an augmentative and alternative communication device is] a voice. Why would you deny someone that?
>
> —Donna, Sam's mother

> We just picked a standard little boy [voice on Proloquo2Go]. ... I think I'm going to change it to a younger voice, because it doesn't sound like I would imagine he would. He doesn't really have a ... voice. It's not like he can talk a little bit.
>
> —Anne, Eric's mother

> Raul has a difficult time with speech. When he can't communicate, he doesn't have a voice. Even though [the iPad is] a speech-generating device, which technically has a "voice" that you can hear, I'm also getting at a more basic form of communication in having the voice, having the ability to communicate to others and initiate dialogue, however that may look like.
>
> —Nina, Raul's mother

Voice is a universally relevant concept yet there is no shared consensus of its meaning, particularly among those who study media and communication.[1] This polysemy is especially palpable in the three introductory quotes. Each parent articulates different understandings of voice in relation to their child's use of the iPad and Proloquo2Go. To Donna, an AAC device is a voice, and a voice is a prosthesis and bodily extension. For Anne, her son performs voice through synthetic speech, and in doing so, sounds like an older child. As Nina views it, voice is a means of self-representation.

Recognizing and listening to children's voices—a primary concern of sociologists of childhood—is a complex undertaking, and particularly so with respect to children with no or limited oral speech.[2] Drawing on Tanja Dreher's notion of a "partial promise of voice," I argue in this chapter that claims to the iPad and Proloquo2Go "giving voice" remain only partially achieved at best.[3] Popular celebratory discourse around these technologies obscures more complicated issues surrounding disability and agency as well as conflicting expressions of children's voices. Such oversimplification can lead to unaddressed concerns in the design of synthetic speech and voice output communication systems along with further inequity in the institutions in which technologies and users are embedded.

Moving forward, understandings of synthetic speech technologies need to be decoupled from utopian and dystopian notions of futuristic conversational agents—or "computer voice." We must also explore the ordinary meanings that humans with significant speech impairments and their conversation partners assign to such voices when used in electronic communication aids. While computer scientists and interaction designers ask questions about how human a talking computer should sound, we need to reckon with the fact that speaking through a computer is also a human way to talk.

Giving Voice to Synthetic Speech

Before delving into more abstract notions of voice, it is important to note the concrete ways in which AAC devices are themselves given voices. A "voice" for nonspeaking individuals through the use of technology does not necessarily involve the electronic production of speech. For example, the Picture Exchange Communication System (PECS), a popular low-tech AAC system, involves children with complex communication needs exchanging laminated paper cards of picture icons with another person in order to ask and answer questions, make comments, and issue requests. PECS cards are stored in a binder, usually attached to Velcro strips. Speech therapists sometimes refer to the PECS binder as the child's "voice," under the premise that this colloquialism socializes caregivers to the idea that the PECS system supports their child's communicative agency.[4] iPads and AAC apps can digitally store thousands of messages in a much more efficient, compact, and archival manner than a PECS binder.

The voices that AAC technologies produce are also never completely disembodied. Ideas about the normative body, vocal techniques, and communicative repertoires have historically influenced the design and development of voice output communication technologies. Long before smartphones could speak multiple languages and GPS devices could spout driving directions, ancient civilizations attempted to mechanically simulate the human instrument of voice. The Greeks and Romans would rig statues with concealed speaking tubes to make their idols appear to talk.[5] Talking automatons designed in the eighteenth century drew inspiration from the *vox humana* pipes of an organ, which imitated a chorus of human voices.[6] The engineers of early speaking machines attempted to copy the vocal organs and model "normal" human physiology.[7]

In the present day, there are two main kinds of voice output used in electronic AAC devices: digitized and synthetic text-to-speech output.[8] Digitized output is any kind of recorded speech or nonlexical sound (e.g., laughter) that can be prerecorded and played back. Text-to-speech output uses software to translate visual text into audible speech.[9] Contemporary techniques for generating synthetic speech include a process known as concatenative speech synthesis in which human speech (usually produced by an actor or actress, but sometimes engineers themselves) is recorded, broken down into units, stored in a database, and recombined into synthesized words. Older systems require more digital signal processing, which causes these synthetic voices to sound less "natural."[10]

While engineered voices at present sound less robotic than early systems, they are not quite human either, a slippage shared by another pervasive technology: online bots that emulate and automate human activity and interaction.[11] One giveaway is that synthetic speech tends to lack prosody—the stress, intonation, and rhythm of human speech. Humor poses another problem. Movie critic Roger Ebert, who lost the ability to produce embodied oral speech following complications from thyroid cancer, famously proposed an "Ebert test" for humor.[12] Invoking Alan Turing's Turing test, Ebert challenged engineers to develop a computer-based synthesized voice that would be indistinguishable from a human one in its ability to tell a joke well. Even the most sophisticated synthetic voices on the market are a far cry from the vibrancy of Samantha, the fictional operating system and titular character in Spike Jonze's 2013 futuristic film *Her*.[13]

Lack of voice customization and dissatisfaction with voice quality can increase the likelihood of an AAC device not being used.[14] Though most high-tech AAC systems offer some sort of choice in synthetic voices, they are rarely personalized. One kind of customization is voice banking, the process of recording one's voice for future use on a communication device as digitized or text-to-speech output. Voice banking is usually done prior to when an individual with a degenerative condition experiences significant speech changes. Alternatively, premade synthetic speech options suggest a narrow range of possible expressions of gender, age, size, race, ethnicity, and regional accent.[15] These voices are not limited to AAC devices, computers, or smartphones; they can be heard in ordinary machines such as cars and household appliances, and ubiquitous text-to-speech like mass transit public address systems and elevator announcements. In an effort to diversify synthetic speech, companies such as VocaliD have turned to crowdsourcing to harvest human voices from willing "speech donors" worldwide.[16] Engineers craft novel and unique voices using samples from a surrogate that closely matches the AAC user, a process akin to organ donation.[17]

For all the technological advancements, some AAC users purposefully choose "nonnatural-sounding" voices at odds with how others might identify them. Stephen Hawking, for instance, declined updates to his synthetic voice because he considers it not only part of his identity but also his trademark.[18] Ebert worked with Scottish company Cereproc to craft a new synthetic voice using samples from his many television and radio recordings as a movie critic and commentator. In his 2011 TED Talk, he mentions dubbing this voice "Roger Jr." Yet Ebert ultimately chose to stick with Apple's "Alex" voice on his laptop, explaining that the "flow isn't natural" using Roger Jr. In sum, the voices of AAC devices do not only convey messages; they signal particular identities along with varied meanings of the essence of human and machine.

The Voices of Proloquo2Go

Ironically, Proloquo2Go (along with augmentative and alternative communication) is quite a mouthful to say. The voices that the app provides mechanically and culturally code identity into the synthetic speech software. Proloquo2Go does not require an Internet connection to operate, but one is needed to download voices beyond the preinstalled ones named "Tracy" ("female adult") and "Ryan" ("male adult"). If the Apple device is not connected to the Internet during the initial setup for Proloquo2Go, then one of these "lower quality substitution [voices] will be used until the app gets a chance to download the voice," as the Proloquo2Go user manual states.[19]

As of June 2016, Proloquo2Go offered users twenty-four voice options that speak "US English" (figure 2.1). Of these, there are sixteen male options, but only nine female ones. In addition to this quantitative discrepancy, there is a qualitative one. The "Will" voice offers a wide range of "expressive

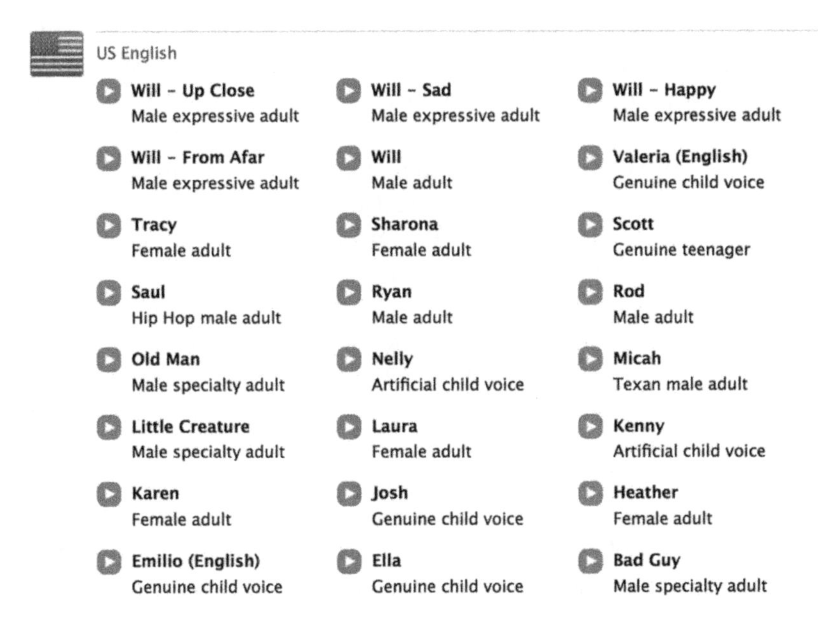

Figure 2.1
Screen grab from AssistiveWare's Proloquo2Go website, listing its available voices for download to use with the app, along with audio samples of each for listening. Image copyright AssistiveWare B.V. Used with permission.

adult" versions—up close, from afar, sad, and happy—but there is no expressive female equivalent. There are also no gender-nonspecific voice options in Proloquo2Go. For example, the neutrally labeled "Little Creature" voice sounds remarkably like *Star Wars'* Yoda. These forms of synthetic speech reproduce gender essentialism and reinforce gender binaries.

Only three of these twenty-four options suggest a nonwhite speaker, and these synthetic voices pose their own problems. Proloquo2Go, for instance, offers "Saul," a "hip-hop male adult" voice, ostensibly coding blackness through the cultural marker of rap music. One AAC specialist in South Los Angeles suggested that her predominantly black and Latino/a/x students were appropriating the Saul voice regardless, noting, "I have a student who uses this to tell, 'Yo, mama' jokes. ... I have some kids who are so motivated to use this [Saul] voice because the app comes with it." While "Saul" may take on new meanings through its use, there is not a single option built into the system associated with an adult female of color who speaks "US English."

Of these voices, Proloquo2Go offers seven for children that were codeveloped with speech technology company Acapela Group. AssistiveWare purchases software licenses from various third-party companies in order to produce Proloquo2Go. This includes synthetic voices from Acapela, a symbol library from SymbolStix, and a grammar engine from Ultralingua. The synthetic speech options in Proloquo2Go include two "artificial child voice" options named "Nelly" and "Kenny," and four "genuine child voice" options. The first of these to be introduced were named "Ella" and "Josh."[20] A behind-the-scenes video on the AssistiveWare YouTube page on the making of the Ella and Josh voices shows that the company used sounds produced by a white boy and girl.[21] Another voice is "Scott," the "genuine teenager." In the absence of a teenage girl voice, however, the genuine teenager is a boy de facto.

The choices made by software engineers and app designers—who are overwhelmingly older white cisgender men—fundamentally shape synthetic expressions of voice through an AAC system like Proloquo2Go. Individuals who use AAC devices ought to be able to use voices that reflect the full kaleidoscope of their identities at every turn, just as all individuals should get to choose which terms (e.g., nicknames, gender pronouns) others use to refer to them.

A Partial Promise of Voice

The social context of these expressions also matters a great deal as to whether or not said voices are listened to with respect. Society tends to privilege oral speech over other forms of communication as a manifestation of truth. This dates back to Socrates, who understood the written word to be a barrier to "pure communication."[22] Speech is often idealized as necessary for "authentic" human connection.[23] In the English language, what it means to be a not fully developed human is actually defined by a deficit of speech; the Latin root of "infant" (*in fans*) means "not speaking."[24] This fetishization of corporeal voice dominated Euro-American technoscience in the 1940s and 1950s, particularly cybernetics and communication engineering.[25] The Turing test, as a case in point, determines humanness based on the ability to converse.[26]

But if we take for granted that everyone has an embodied voice, then we run the risk of disenfranchising individuals who do not or choose not to communicate through oral speech. People with disabilities are frequently spoken *for* without having a say of their own.[27] The voices of those children and adults considered "nonverbal" are particularly discounted.[28] While democracies claim to represent *vox populi* (the "voice of the people"), ordinary citizens—including nonspeaking and minimally speaking individuals—struggle to be heard through traditional and new communication tools.[29]

Dreher offers a useful way of conceptualizing voice that acknowledges tensions between uncritical celebrations of "speaking up" and the importance of having a "voice that matters."[30] She uses the phrase "a partial promise of voice" to highlight the responsibility that social institutions bear to follow through on aims to promote the voices of marginalized groups with commitments to response and recognition. Speaking and listening are intertwined, writes Dreher, forming "circuits of democratic communication" that frequently breakdown.[31]

While Dreher focuses on the dynamics of listening at a policy level, the notion of voice as something dynamic and distributed has also been taken up by critical scholars of speech impairment, society, and technology.[32] Voice is a multidimensional social construct; the voices of nonspeaking children are not static entities to be found but rather dialogic processes.[33] The ethics of listening to disabled publics is crucial to consider, too, in any discussion of voice and speaking.[34] Technologies create new speaking as well as new listening practices in relation to disability (e.g., hearing aids and cochlear implants), but narrow conceptions of normal speech and hearing embedded in culture, democracy, and media can compound injustice and discrimination. "Natural" and "unnatural" voices are socially constructed categories.

Building on this work, the following sections loop parental understandings of their children's voice—conceived by them as something that is at once embodied, performative, and political—back into liberation narratives about the iPad and Proloquo2Go. Popular discourse about screen-mediated communication does not address the overall inequalities in how nonspeaking individuals' voices are valued, or the unequal distribution of technologized voice across gender, race, ethnicity, class, and linguistic background. At present, the promise of voice through the iPad and Proloquo2Go is only partially fulfilled.

Voice as Embodied

One way of understanding voice is as a medium that extends the communicative capacity of the body over a distance.[35] Leopoldina Fortunati writes, "Given that the voice is an extension of the body, we could say that the body reaches where the voice does, that is, a good many meters away from the body itself."[36] There were two ways in which parents understood the possibilities and limitations of their children's voices pertaining to the human body: tensions related to one versus many bodies in the embodiment of voice, and those related to technology and the human in vocal embodiment.

Body and Bodies

Whether voice is an individual or social construct is a largely unsettled issue in our culture writ large. A key friction emerged among parents as to whether or not any individual besides the child AAC user (e.g., other children and peers, or parents themselves) should be allowed to interact with the iPad and Proloquo2Go. Parents and professionals agreed in principle that a nonspeaking child's AAC system should always be available to them. But unlike vocal organs housed inside one's body, another individual could potentially use the child's iPad and Proloquo2Go to produce speech. This led to debate around whether voice resided solely in the child's body or was distributed across multiple bodies supporting the child's synthetic speech production—particularly the bodies of mothers.

Children and Peers Some parents were concerned that other children, especially those unfamiliar with the concept of AAC, might view the iPad with Proloquo2Go as a toy. Though the iPad is often praised as an AAC device due to its social acceptability, this desirability can also interfere with its function as a personal speech aid.[37] For example, when I asked David if Beatriz brought her iPad along when visiting extended family, he replied, "No, because then kids, based on her condition, kids take it away from her, go play the games." The cool gadget made Beatriz a target of other children's desires for the luxury good, and it was challenging to defend herself against those jealousies without adult intervention.

Besides child family members, some parents had anxieties about classmates and peers wanting to use the iPad. Nina spoke of Raul's young classmates, "Sometimes they want to play with it and they think it's a game because that's what they know an iPad to be. But then I have to say, 'This is Raul's voice, this is how he uses it.'" The technology's mixed use created both opportunities and limitations in certain social contexts. Alice reported that the kids at church liked to play with Danny's iPad. She didn't mind if they wanted to use it, "but he needs to be able to speak."

Some nonspeaking children were able to assert limits on their classmates' curiosity. Peter reported that Danny used AAC to resist his classmates' attempt at using the iPad without his permission. "Some kid came over to play with his iPad," said Peter, "and [Danny's] like, 'Stop, stop, stop.' Then the kid stopped, and he was like, 'Go, go, go, go, go. Go away.'" Rebecca noticed that Kevin had accidentally taken a video with the iPad that documented other kids on the playground trying to use it. "He was blocking so the kids couldn't play with it or whatever," Rebecca remarked. "But he hung onto it, and the other kids were respectful. They didn't take it away from him or anything." Children and peers covetous of the iPad shaped how and if Beatriz, Danny, and Kevin expressed themselves through the technology.

Parent and Child In addition to other young people, some parents were concerned about interfering in their child's communication by using the child's iPad and Proloquo2Go to speak on *their own* behalf. Perri was initially using Cory's iPad to talk to him, but later stopped. She explained that when Cory's speech-language pathologist encouraged him to communicate using Proloquo2Go, the therapist said to him, "Where's your words? These are your words. Good using your words." Perri interpreted this to mean that the speech produced by Cory's iPad were his words and his alone. She decided then to separate "Mommy's words" from "Cory's words" by downloading Proloquo2Go onto her own iPad, and modeling use for Cory on her tablet. This meant, though, that she carried around two iPads—hers and Cory's—at all times.

Other parents embraced the fluidity between themselves, their child, and the child's iPad—a liminal state in terms of where one person's body or voice ended and another's began. Daisy, for instance, felt that there were some situations in which it was better if she talked through Thomas's iPad instead of using her own embodied oral speech. In Thomas's Proloquo2Go, there was a folder for "Mommy's words" that included phrases such as "No biting," "No grabbing," and "No kicking." Daisy reported that Thomas "wouldn't listen much to me when *I'm* saying it, but he would listen to the iPad. When he started [having a] tantrum ...and biting, he would listen to this. ... I just keep on hitting this over and over, then he stops." Daisy had her own theory as to why Thomas obeyed his mother's instructions when spoken through his own iPad. "I guess it's the tone, computerized ... the sound of it," she said. Her voice and the synthetic voice of Proloquo2Go symbolically blended together at times.

Nina thought that it was important for her to model the use of Proloquo2Go for Raul. She remarked that the behavioral therapists that initially worked with Raul told her, "That's his [iPad]. He touches it, and nobody else touches it, and that's it. We don't use it." Nina began to do some online research, though, and came to a different conclusion. "It was like, 'Hey, we need to use this too because it's like learning a second language.' You have to model that language for them." Nina understood the iPad as both belonging to Raul and distributed across the adults supporting his use of it. "So now I see it more like, OK, it's not just *Raul's*. It's like, it *is* Raul's, but we need to use it to interact with him," observed Nina. A mother's touch guided Raul's interaction with the screen and his production of speech.

There was a clear gendered element to this distribution of voice. For Nash (age three), who had limited motor control due to cerebral palsy, his mother's body became a key component of his communication system. Because he used his full body to gesture and touch the iPad, Nash had difficulty using an iPad mounted on his walker or wheelchair because the physical mobility system restricted his upper-body movement. Instead, I watched as his mom, Taylor, nimbly held Nash and his iPad simultaneously, while he shifted his body weight in her lap and extended one arm to select a message from the screen. Taylor became a more responsive seating system and iPad mount than the inert objects. Yet this interdependency was not distributed equally between Nash's parents. Rachel asked Nash's dad, Todd, if he ever used Proloquo2Go with Nash. Todd said that he didn't, and Rachel asked why. Compared to Taylor, he shrugged and replied, "I feel like I'm not holding him right, or the iPad right."

Mothers in particular tended to physically bear responsibility for their child's communication through the iPad. Nina felt frustrated when she forgot to take Raul's iPad on a recent family trip to the pumpkin patch. "I usually bring it, or I have him bring it with us," explained Nina. "I told [Raul], 'I'm sorry, buddy. I forgot your talker. Well, we'll just hang out here.'" Kameelah reflected on the affective labor involved in being a bodily conduit for Talen's communication. "I feel *horrible* when we leave the house and we forget the black iPad," she said, "because I'm like, 'How would I feel if somebody just like ripped my voice box out?' Because, you know, nobody holds my voice but me, right?" Kameelah felt personally responsible for impairing Talen, likening forgetting his iPad to leaving behind a part of his body. "You're extra handicapping him or extra disabling him by [leaving the iPad at home]. I feel really bad when that happens," she disclosed. In this way, it was a mother's job to be fully prepared for her child's communication attempts, and prevent them from becoming voiceless without immediate access to mobile speech technology.

In addition to the "second shift" work of household labor, US mothers are largely charged with decision-making tasks regarding their children's media and technology use.[38] This dynamic process involves the additional "third shift" work of managing their own emotions and anxieties about both parenting and their children.[39] Perri, Daisy, Taylor, and Kameelah balance their desires to be good mothers with the daily demands of supporting their child's well-being and use of assistive technology. While the voices of children using the iPad and Proloquo2Go are singular as well as mutually constructed—shaped by social interaction with siblings, peers, and parents—the bodily distribution of both voice and voicelessness is heavily weighted toward mothers.

Technological and Human

Besides individual and collective notions of embodied voice, another tension emerged between the technological and human, and how voice is owned, controlled, and accessed across these mutually shaped domains. This friction manifested in three main ways. First, parents used different terms to refer to the iPad, including describing the device as a prosthetic "talker." Second, they debated whether the voice produced by the iPad had a positive or negative impact on the child's embodied oral speech production. Third, some believed that technology might provide a more direct connection between the child's "inner voice" and their speaking voice. In these different understandings of the embodiment of voice, class, cultural, racial, and ethnic distinctions emerged.

Nomenclature More and less privileged parents differed in the terminology they used to refer to their child's iPad with Proloquo2Go. A number of more privileged parents and professionals called the child's iPad a "talker," whereas less privileged parents tended to use more generic terms like the "iPad" or "device." Nina noted, "I had heard other people say 'talker,' so we just picked that." Rachel, one of the speech-language pathologists I shadowed, encouraged a number of her clients to use the term. "What do you call it?" Rachel said to James's mom, Cathy, holding up the iPad. "Your words, your device, the iPad," Cathy offered. "It's not that we try to teach him to say 'device,' but we say it when we are with other people." Rachel recommended Cathy call the iPad a talker instead. "It's just a way to call ... whatever this is to you," she said suggestively. "It's just that with calling it 'iPad,' he might think that it was for other things, like games, and not for talking, which is what it should primarily be for." This exchange highlights how sociotechnical systems can reinforce class distinctions due to institutionalized cultural capital. In the view of speech-language professionals whom I observed and interviewed, there were "right" and "wrong" names to call the iPad with Proloquo2Go. More privileged parents tended to get it right, while less privileged parents were regularly doing it wrong.

The term "talker" encompassed a range of meanings for more privileged parents: a tool that a person uses for talking, a machine that produces speech outside one's own body, and a term for older AAC devices. During Danny's applied behavior analysis therapy, Alice and Peter would leave one iPad "as his talker and the other one can be a reward," Peter explained, "so you're not taking away his voice when you're taking away his iPad." Alice and Peter also used talker to refer to more traditional, bulkier AAC devices. "Have you picked up a talker?" Peter asked me. "It's not portable for a little child." Alice also drew a distinction, noting that an iPad was "more socially acceptable than a talker."

Less privileged parents used nomenclature for the iPad that was less explicitly associated with talking. Kameelah referred to Talen's AAC device as the "black iPad" and was reluctant to call it a "voice." She had encountered a speech-language pathologist that used the latter term. "She would call it, for her kids, 'Where's your voice? Don't forget your voice. Bring your voice.' I don't know how I feel about that." While Perri embraced the iPad as "Cory's words," Kameelah was averse to adopting the professional language. She felt that signifying the iPad with the word "voice" implied that Talen would be essentially voiceless if he were without the device. "I don't want you to feel like you can't communicate if you don't have it," she explained, imagining what she would say to Talen. Kameelah thought it was important to decouple the machine from the body, although she hadn't decided on a permanent way to refer to Talen's iPad with Proloquo2Go. "It's got to be the perfect word that also leaves it open for, 'This may not be something that you use forever,'" Kameelah said. The tethering of machine and body could be temporary and tenuous, and rigid professional terminology around the iPad and Proloquo2Go did not reflect this fluidity and liminality.

Embodied Oral Speech In terms of the coupling and decoupling of technologies and bodies, parents also discussed whether speech output produced by the iPad would promote or prevent their child from talking. Before acquiring the iPad and Proloquo2Go, some initially had faith that their child would eventually produce embodied oral speech. Daisy admitted that when Thomas was in first grade, she was "still hopeful that he's gonna talk." She thought at the time, "No, speech therapy. Let's teach him how to talk. I don't want him to use that device, 'cause he's not gonna ever learn to talk. And he's just gonna be dependent on that device." Donna similarly described her initial thinking as "'No, no, no, he'll be OK. He just needs a little bit longer.' I was a little bit in denial about him needing a device." Donna and Daisy deferred introducing their sons to a voice output communication aid, hoping that with time and speech therapy, they would talk.

For parents whose children had initially started talking when they were younger, the introduction of an AAC device meant letting go of the idea that their child's embodied oral speech would return. Perri, for instance, rebuffed her son Cory's use of an AAC device at first, thinking it was an unnatural way of communicating. "I want his sounds back. I don't want him to tap an iPad," she reflected. Sara said that initially she "didn't want to give up on the thought that [Isaac] was going to use his voice, fully use his voice. In my mind I was thinking, 'Well, if it's going to replace it, he's just going to rely on that.'" For Karun, "'Til now, I [had] hope that [Pargev] might suddenly just talk sentences. I prefer the words rather than the iPad." Rob actively encouraged Luke to communicate in more "traditional" ways. "I'd rather be able to get *him* to speak it than have to use the machine to speak it," he said. "If he can find another way to communicate without it that's more traditional, then we encourage that."

Despite these hesitations, parents described an ongoing process of coming to terms with their child's speech abilities and disabilities. "I came to the realization because [Thomas] was gonna be eleven, that he's not really ever gonna talk," Daisy remarked, and "if he does, it'll just be sounds." After speech therapy and sign language interventions were unsuccessful, Perri also turned a corner. She said, "I'm like, 'I don't care. At this point, I don't care. He has a lot to communicate. I don't know what he's got in his brain, but I know it's a lot. I just want to hear it.'" Cory took to the iPad quickly, which Perri interpreted as, "He was telling us, 'Hey, I want to talk. I need some way to do it.'"

A number of parents observed that their child's use of the speech-generating device was actually contributing to increased attempts at spoken language, which growing research supports.[40] "Now, [Thomas is] making more sounds," said Daisy. "He's copying." Vanessa commented, "I honestly feel like Proloquo2Go has improved [Moira's] verbal language. ... She'll even type it out and say it on Proloquo2Go, and then she'll say it." Voice is not only coconstituted by the child's body and the bodies of others but also in part by the interconnectedness of humans and machines through the rehearsal of communication practices.

The Brain–Voice Connection Parents discussed technologies not only as extensions of their child's orally produced speech but also as the emotions and expressions in the child's mind. Human communication is said to have an "interiority of sound" in that our interior thoughts are private until expressed.[41] Some parents hoped that the iPad and Proloquo2Go would give them access to their child's "inner voice," despite the limitations of their oral speech. Rebecca, for example, said that in the future, she'd like Kevin "to actually communicate more of what he's thinking. ... He might think it in his head but he might not be able to verbalize it. There's the brain–voice connection or something." Nina also expressed a desire for her child to share his inner world with her. Said Nina, "When I say 'voice,' it doesn't have to do with always *hearing* it necessarily, but what's coming from inside them—their emotions, their thoughts, their insights." These parents viewed technology as a way to scaffold connections between the brain and voice.

A number of parents looked forward to a day when technology would provide a more seamless link between the child's inner thoughts and their outward production of speech, like a kind of mind reading. "Probably in the coming years, they are going to come up with something much better," noted David. He mentioned one such innovation, not identifying its source. "There is this device that is interesting me a lot. It is like a little band," he said, gesturing to his head, "and this band interacts with your brain just by looking at the screen ... instead of using your finger."

David could likely have been alluding to an emerging communication technology known as a brain–machine interface. While invasive brain–machine interface technologies require surgery, noninvasive ones record brain signals using electrodes placed on the surface of the head. Even though much of the hype around these technologies has envisioned able-bodied individuals gaining Jedi-Master-like abilities (based on the *Star Wars* films), parents of nonspeaking children also report hope in the technology affording their child with new opportunities to express themselves, communicate their needs and preferences, and expand their social circle of conversation partners.[42] It is unclear, however, if the neurological models on which these brain–machine interfaces are built reflect the full range of human neurodiversity and neurodivergent thinking.

Wearable technologies offered parents additional promises of easier communication with their children. Peter, for instance, mentioned his hope that Danny would one day be able to wear an AAC device. "I'd like to see—and I know it's coming someday—where the technology is a lot more integrated," said Peter. "So we didn't have to worry about carrying it along, so it could be just on a sleeve or something like that." The advancements described by David and Peter would overcome bodily limitations that prevented brain-voice synchronization.

A few parents also made suggestions for novel interactive technologies that might better elicit voice from their minimally speaking children. "Out there, there are games for verbalizing," mentioned Karun, "like the [Microsoft] Xbox games, something that a real person is talking." She imagined a game featuring a "character that the child chooses to be. The character talks instead of him, or something like that." Karun brought up similar products on the market. "Maybe you should track down one of the Xbox games that my son plays," she told me. "It has people who talk. They don't talk a lot. One of them, which is Sonic," she said referring to the title character of Sega's *Sonic the Hedgehog* franchise. "Sonic talks a lot. Something like that for autistic children."

Karun is not alone among parents of children on the autism spectrum in suggesting that interactive virtual agents, which are increasingly embedded in ordinary mobile communication technologies, might solicit speech. Research on neurotypically developing children and their interactions with networked voice input on mobile devices suggests that they use systems like Apple's Siri and Microsoft's Cortana, which are designed to recognize adult speech patterns, for a variety of aims. This includes exploring and understanding the nature of interactive agents, information seeking online, and operating the device hands free.[43] Still, the research is unclear as to whether or not minimally speaking children on the autism spectrum might differ in their uses and intentions for interacting with voice input systems. There is also no consensus in the human–computer interaction literature regarding whether animated and/or nonhuman forms are more effective than humans at eliciting responses.[44]

Users have the opportunity to form intimate relationships with disembodied assistants such as Siri or anthropomorphic characters such as Talking Tom Cat from the eponymous app.[45] Understanding these relationships requires situating such interactions within individuals' social contexts and existing patterns for how they express their sociality, which can take myriad forms for autistic youth with complex communication needs. Nelson, for instance, contended that the Talking Tom Cat app offered Stephanie (age ten) and other autistic children "a creature that repeats everything they say and has the patience to listen." The interface interprets her attempts at speech, recognizes it as input regardless of the vocal quality, and in turn presents an entertaining output. Using her voice, Stephanie can feel powerful and is incentivized to elicit reactions from Talking Tom Cat.

Nelson explained, "She makes a kitten meow, she makes [Talking Tom Cat] jump up on the ceiling, she makes him fart, she knocks him down." Marisa mentioned that Stephanie's vocal interaction with the Talking Tom Cat app was the first time "we really hear[d] her voice." Karun, Nelson, and Marisa described a permeable distinction between inner and outer voice—one constantly being created through interactions with other people and responsive machines.

There nevertheless are significant issues regarding which parents are able to actually leverage these interactive experiences into improved opportunities for their children. In October 2014, the *New York Times* published an article titled, "To Siri, with Love: How One Boy with Autism Became BFF with Apple's Siri."[46] In the article, Judith Newman, a parenting blogger, describes how Siri motivates her autistic son, Gus, to communicate. Newman writes, "Gus speaks as if he has marbles in his mouth, but if he wants to get the right response from Siri, he must enunciate clearly." His bond with Siri is forged because "she" reliably and patiently retrieves information about his specific interests in airplanes and turtles. Moreover, Newman claims that her mother–son relationship with Gus is actually being changed for the better by Siri.

Although Gus and Stephanie might use speech agents to support their communication in similar ways, their backgrounds could not be more different. Gus's mother is a white, Ivy-educated, tech-savvy parenting columnist for one of the world's most preeminent newspapers. Stephanie's mother, Marisa, is a Latina immigrant from Mexico and nonnative English speaker without a college degree. Gus attends a state-of-the-art private school for autistic children. Stephanie's parents, on the other hand, feel that they receive little support from public school professionals and therapy providers. Marisa lamented, "They showed us how to open Windows and all that, the basic thing, but not how to use [the iPad and Proloquo2Go] for helping Stephanie or [us] understand what we can do with the device." Parents' ability to mobilize social, cultural, and economic capital shapes how helpful these widely accessible tools can be in reality.

While artificially intelligent voice response systems and chatbots invite utopian and dystopian visions, as being either culturally liberating or "socially destructive," it is critical that we not theorize about these tools divorced from their ordinary use.[47] Among parents, there was no uniform understanding of the embodiment of voice, both in terms of whether voice was an individual and/or social construct, and whether or not technology could be a compensation and bodily extension of voice.

Voice as Performative

In addition to an embodied component, parents discussed voice as a means for their child to perform identity. In this sense, identity is not a static or single entity but rather a set of tensions shaped through interactions with one's environment.[48] Parents talked about the synthetic speech generated by the iPad and Proloquo2Go as having to do with four aspects of identity performance: sounding one's age, sounding human, sounding related to one's family members, and sounding of one's cultural background. Beatriz's case illustrates the complicated entanglement of these various aspects of identity and performance through voice when coming into contact with different sociotechnical systems that privilege dominant groups, namely white cisgender males.

Age

A number of parents wanted their children to speak with a synthetic voice that reflected the maturation level of their child. Alice noted, "I know we wanted a child because it would just be weird for Danny's voice to be like an adult male. It would just be odd. Because you want it to fit your kid." Esosa said that Chike's school had initially set his voice in Proloquo2Go to sound like an adult man, which at the time was the only male option. She remarked, "I don't want it to sound like a man." When Proloquo2Go released child voices, "I picked Josh, who's a boy," Esosa added. Donna meanwhile was preparing for the day when Sam hits puberty and no longer needs a child voice. Said Donna, "Once Sam's voice changes and he gets to that age, then we'll switch the voice to a more mature voice." Childhood and adulthood along with their relationship to gender are vocally performed in part through the machine's synthetic speech offerings.

Human Expression

Not only did some parents value a voice that sounded like someone of the same biological age, but they also preferred one that sounded like a "human" child. Moira's mother, Vanessa, explained that Proloquo2Go "came out with an updated version and they had a better kid voice. ... The one [Moira] has now definitely sounds like a little girl." Similarly, Perri noted, "I found the voice that sounded most like a human little boy." Others did not mind if an AAC device provided a less natural-seeming voice, similar to Hawking's and Ebert's preferences. "The voice that comes out of that, that's his voice to me," insisted Donna. "That's Sam. I think the voice is called 'Kenny.'" Donna said that she liked the "quality of the voice that comes out" even though "there is a tiny touch of roboticism to it." To Donna, the nonhuman "Kenny" voice took on the human qualities of Sam, not the other way around. "Whenever I hear Kenny in the machine speak," she explained, "that is Sam to me." What it means to sound human is performed through, not in spite of, mobile technology and synthetic speech.

Kinship

In addition to desiring a synthetic voice that mimicked the vocal qualities of a human child or assimilating synthetic speech into perceptions of the human voice, some parents wanted a voice for their child that sounded biologically related to their family. "If Danny spoke, he would sound much like me at that age, right?" Peter asked rhetorically. Unaware of the efforts of companies such as VocaliD, Peter wished "they would allow you to record some subset of syllables or whatever, and then pitch it down, make it higher pitched, and say, 'Hey, there's Danny's voice.'" Peter thought it would be ideal for Danny to have a customized sound "because most people in families sound about the same, pitch about the same, cadence about the same." Peter wanted the design and development of synthetic speech to more closely mirror biological reproduction and family socialization.

Some parents, though, chose voices for their children that purposely did not *sound* related to them but instead connected them in other ways to their idea of family. Mark commented that they had chosen River's voice on Proloquo2Go largely because it bore the name of River's relative. "He's got a cousin Liam, so that was kind of cool. ... We checked out 'Liam' because it's his cousin's name, and it just seemed like the right cadence and everything," Mark said. Proloquo2Go categorizes the "Liam" voice, though, as "Australian English," and neither River nor his immediate family members are from Australia. Children perform family belonging and identity through their synthetic voices, but not always in an expected manner.

Culture

What it means for family members to "sound about the same" is about much more than biology; cultural identity can be as, if not more, important. Esosa remarked, for instance, that it was "good" that Proloquo2Go had released a voice that sounded like "an Indian kid speaking English." Although she herself was originally from Kenya and not India, she observed, "If you're from an Indian family and everybody sounds a certain way, you can sound the same way if you want." As of June 2016, however, there was no Indian child voice in Proloquo2Go; the only available "Indian English" option was for "Deepa," a "female adult." Things were even more complicated for AAC users who identify as Indian American. The "Deepa" voice appeared next to an icon of the Indian flag; of the voices marketed in Proloquo2Go under the category of "US English," there were none with an Indian accent or traditional Indian name.

Not all US children had the same options to sound like their family members, even if they wanted to. One AAC consultant mentioned that there were "hundreds of languages spoken in these schools" in Los Angeles, but her district could only provide systems in English because it was the language primarily spoken in schools. This created an inherent communication barrier between home and school cultures when the AAC system went home to families. She noted, "One of the kids I work with here, his parents speak Korean at home. Any kind of assistive communication system, they wouldn't use it because they don't speak it." Proloquo2Go has no Korean-accented, English-speaking voice options, nor a version with vocabulary in Korean.

The app also struggled with pronouncing names and terms that originate in languages other than English. Sara explained that "there's all the Jewish food [Issac] eats, like challah. I typed it in [to Proloquo2Go]. It doesn't say it the right way, because they can't say the 'C-H' the same." Issac's voice was rendered distinct from members of his family and Jewish community because the software did not recognize such variation, bounding his potential performance of culture and identity.

Not According to Her

Linking together the themes discussed above, Beatriz (age ten), a Latina girl with cerebral palsy and epilepsy, performed age, human expression, kinship, and culture in various ways through her voice. While her developmental disability impeded her ability to produce embodied oral speech, I would still call her chatty. When I met her, she knew a few signs in American Sign Language, like "water" and "cookie," and used these to communicate at home with her family. In her own way, she also conveyed phrases like "Oh, me!" uttered orally with a blend of sass and exasperation that reminded me of the preteen girl characters on the Nickelodeon and Disney Channel television shows she liked so much.

The computerized voice that Beatriz used with Proloquo2Go did not impart her personality, though. Prior to using the app on the iPad, Beatriz's high-tech AAC system consisted of an Apple iPod Touch, an Amazon Kindle, and another app called TapToTalk installed on both. Unlike Proloquo2Go, which employs synthetic speech, TapToTalk uses digitized speech. The app allows someone to speak into their mobile device's microphone and record human speech that can then be reproduced when the user taps a corresponding button on the app. Beatriz's older sister, Ariana, supplied a donor voice for Beatriz to speak with through TapToTalk, a DIY "hack" of the reverse-engineered familial voice that Peter described above.

Beatriz's father, David, remarked that his family preferred the prerecorded sounds of Ariana to the synthetic speech of Proloquo2Go. He explained that there were three main reasons for this preference: "because it was her sister's voice, and it was according to her age, and it was on the tone of voice that we wanted." David's desire for Beatriz to speak through her sister's voice reflected the bonds of affection between the girls and their family. As I was interviewing David and Pilar, Ariana came home from school. Beatriz jumped up and bounded across the room to hug Ariana tightly. Her parents and older sister laughed over this warm display of affection. "There is a lot of love around this house," David pronounced happily.

In addition to reflecting their relationship, Ariana's thirteen-year-old voice sounded close in age to that of Beatriz. Of the Proloquo2Go synthetic speech, David said, "It's like an *old woman* or an *old man* talking." Besides sounding naturally related to Beatriz and similarly youthful, Ariana used more expression in her phrasing than Beatriz's current Proloquo2Go voice. "Sometimes, it sounds too automated," protested David. He reenacted for me how his older daughter spoke when recording her voice for Beatriz's use—"I want bacon! Water, please!" He recalled, "I remember when her sister recorded phrases on the other program. … She was putting *feeling* into it. Because you were customizing the phrases."

Since my initial interview with David and Pilar in February 2013, AssistiveWare has made some progress in presenting more expressive vocal options. The release of Proloquo2Go 3.0 in May 2013 included a new "ExpressivePower" feature that allows child users to say emotive phrases (e.g., "It's not fair!") and sounds that children might make during play (e.g., a dog bark). But each time that Beatriz used her sister's voice in TapToTalk, there was an additional personal touch to every button tapped. The Proloquo2Go voice was "not according to her" because it was robotic, not a fellow Latina in their family, and that of an older woman.

If the family preferred the TapToTalk voice to Proloquo2Go, then why not use TapToTalk instead? David explained that it wasn't so simple. After the school district provided Beatriz with the iPad and Proloquo2Go, the family "decided not to continue paying for the [TapToTalk] annual subscription, because she cannot go using two—one here, one there." David thought that Beatriz using a singular voice, one that the school district would freely support her using at school, was more important than Beatriz speaking with a voice that had a better all-around fit with aspects of her identity. "I know it's *good*," he said of the Proloquo2Go software, "but we just need to find ways of ... changing the tone of voice first."

In January 2014, AssistiveWare and Acapela Group made some strides in diversifying synthetic voices for US children by introducing "Valeria" and "Emilio," the "World's First Bilingual Spanish–English Children's Text to Speech Voices."[49] Initially, "Valeria" and "Emilio" were only made available for AssistiveWare's visual storytelling app Pictello, text-based communication app Proloquo4Text, and Infovox iVox voices for Mac systems, but by December 2014, they were available for download on Proloquo2Go. The app did not actually support bilingual functionality between Spanish and English until September 2015, when the company announced that it "now gives a voice to Spanish children" (though ostensibly also Spanish-speaking US children who are not Spanish citizens).[50]

AssistiveWare still has a long way to go in terms of offering voices that reflect a wide range of cultural and ethnic backgrounds, particularly in the US. The story of Beatriz and her family illustrates the need for even more extensive translation work so that individuals with speech disabilities can perform voice in harmony with their own evolving identities, including at the intersections of their gender, age, and ethnicity. Synthetic speech options or the lack thereof made a difference in parents' understandings of their child's voice. More privileged parents of white boys, like Perri and Donna, could more easily identify their sons in the default voice settings on Proloquo2Go and envision the synthetic voice maturing along with their child. Less privileged parents, particularly parents of nonwhite girls, had more difficulty in hearing their child in Proloquo2Go's voice options. This discussion of voice output illuminates the ways in which social and technical systems can advantage children with the right kind of embodied cultural capital with respect to the manner in which a person speaks. Proloquo2Go and synthetic speech systems in general privilege the literal voices of white men over nonwhite women.

Voice as Political

Designer Graham Pullin writes that "communication aids are not a neutral technology or transparent medium"; they are inherently political, for they make visible or invisible certain people, values, and ideas.[51] As such, parents discussed their child's voice as being political in the sense that the iPad and Proloquo2Go were essential tools for the child to exert their own agency. These tools were at times not available to their child to do so. Nina stressed that "without the iPad, [Raul] has no voice." This empowered and yet disempowered notion of voice was explored in three situations by parents: when the child's safety was at risk, when the child was experiencing physical pain, or when the child was articulating resistance to something in their environment. In addition, both parents and professionals mentioned the role of data logs stored within the Recents View section of Proloquo2Go as a form of inherently political surveillance or countersurveillance.

Safety

Parents understood the iPad as something children could use to help themselves stay safe. Mark explained, "If [River] gets lost in the mall, he's got his device. He could just be like, 'This is my dad's phone number, call him.'" Some parents also remarked that the ability to use one's AAC device in an unsafe situation was impacted by whether or not law enforcement was prepared to deal with their child's disability. Due to this uncertainty, Peter was teaching Danny to be able to gesture "yes" and "no" so that he could "at least answer twenty questions." He added, "If a cop's asking [Danny] questions, and got a gun on him, no cop in the world's going to allow him to grab a talker." While Danny is white, being black or brown further increases the dangers of encountering law enforcement officers while being nonspeaking and disabled.

Moira's mother, Vanessa, relayed a harrowing story having to do with these concerns: repeating one's phone number, getting lost, and encountering law enforcement as a nonspeaking minor. It is a story that deeply complicates celebratory narratives about access to the iPad and Proloquo2Go as equivalent to having a voice. Unlike most of the Proloquo2Go users I spent time with, Moira directly input all of the words that her iPad spoke by using the app's Typing View. Vanessa said that she was generally happy with Proloquo2Go, except for its pronunciation of numbers. "I'm trying to teach Moira to memorize phone numbers," she told me. Vanessa explained that one digit at a time, Moira would type in "6-5-7 or 7-1-4, but it'll be like, '714,' '992,' '0008.'"[52] Moira tried to work around this by spacing the numbers out, but found it to be awkward to type. "Teaching her to have to put those spaces in between is so weird," Vanessa lamented.

Teaching children to memorize their own phone number and repeat it to a helpful adult in case of an emergency is typically a top priority for parents and teachers of young children. Vanessa had cause to be more motivated than most, though. "The reason why I'm trying to teach her this," she reported, her voice shaky and with tears welling in her eyes, "is because three years ago, I woke up one morning and she was gone. Front door was wide open. I ran outside. I literally pissed myself. All I saw was white light. I was just in shock." Moira had walked to Starbucks, one of her favorite places, but "didn't take her iPad and went in her nightgown, no shoes, no undies, nothing—I'm thankful she even had a nightgown on," Vanessa spoke with a shudder.

Outside her home, and without her mother or her iPad, Moira encountered a situation in which no one around knew how to communicate with her. The employees at Starbucks, Vanessa recounted, "were like, 'Who does this child belong to?' First they thought she was Deaf and then they realized she was responding to them. They're like, 'She must have autism.'" The Starbucks employees luckily then called the local police, who picked Moira up and brought her to the station. They "tried to get her to write," said Vanessa, but "her handwriting's chicken scratch." It was a nightmarish situation: Moira had no effective way to communicate, and her mother had no idea where she was.

After discovering that Moira was gone, Vanessa picked up her own phone and dialed 911. "They were like, 'We have her. She's here and she's safe,'" Vanessa recalled. Once at the police station, she was dissatisfied with how the officers had handled the situation. She told them that Moira types to communicate. "Then they're like, 'Ohhh! Ahhh! We didn't even think of that!'" Vanessa remarked. The officers lacked awareness of how else a nonspeaking individual might communicate besides oral speech or written text. "It seems like there needs to be more training in that, and there are police agencies that do have that training available. But clearly our police station didn't," Vanessa pointed out, with thinly veiled resentment.

The police department would likely have had an easier time identifying Moira and contacting Vanessa if Moira had been carrying an ID. Vanessa, however, explained that Moira "refuses to wear a medical bracelet. The [occupational therapist's] trying to work with her on wearing it, and it says, 'My phone number,' and then it says, 'Types to communicate.'" Without any certainty that the local police force would get better training or that Moira would eventually be OK with wearing her medical bracelet, Vanessa took matters literally and figuratively into her own hands by teaching Moira how to type her phone number "so that if someone's like, 'What is your phone number?' she can just type it." Vanessa hoped that Moira could use Proloquo2Go to speak on her own behalf in an emergency, but the app's vocal mispronunciation of telephone numbers was hindering her ability to do so.

Pain

Parents also struggled to know when their nonspeaking child was experiencing pain, and hoped that children could one day voice their own discomfort through the iPad and Proloquo2Go. "[Talen] can't tell me he has a stomach ache," Kameelah explained. "He cognitively has to get and understand that when my stomach hurts and they ask me how I feel, this is the icon that indicates how I feel. That is hard work." Since Stephanie was having a difficult time mastering Proloquo2Go, Marisa hoped that her teachers might also support her burgeoning typing skills in case they came in handy. "Maybe Stephanie will be able to tell me, 'Mom, I don't feel good,' by typing," she said.

Some parents identified times when their children were in fact able to use the iPad to successfully relay their physical state. On top of Danny not being able to speak, Alice mentioned, "He has a very high tolerance for pain so it's really hard to tell when he's sick. He certainly doesn't have his body parts down so it's hard to know if a particular part of him [hurts]." Without speech, "the only way we know [he's sick] is he's either vomiting or got a fever," Peter said. Yet "he will try with the iPad to use some novel language to let us know," Alice noted. She told a story about how the last time that Danny went to the hospital to get a regular procedure and came out of the anesthesia, "he did just everything he could to tell us he wanted to go home. 'Off. Out. Leave. Go.' ... He started every kind of word to say, 'Get me out of here,' basically." It is of vital importance for children to be able to voice a pain that no other body can feel, illustrating the limits of an entirely distributed theory of voice.

Resistance

Parents also discussed their children using Proloquo2Go to be able to say that they wanted to pause while doing an activity, wanted to be left alone, and were displeased about something in their environment.

I Want a Break A few parents talked about their children using the iPad and Proloquo2Go to indicate that they needed a break from a social situation. Sara said that she and her family had taken Isaac's iPad with them on a recent trip to Disneyland. "He had a moment where it was just too much and he could tell me, 'I want a break.' It was awesome, because it cut down on the tantrums, and the screaming, and the crying," she explained.

According to Kameelah, Talen was "not there yet" when it came to using his iPad to initiate taking a break. "Now, he just screams," she said. She thought that Talen struggled cognitively with "understanding, in himself, this is where I'm at and this is how I communicate. OK, I'm feeling frustrated. When I'm feeling frustrated, I say, 'I want a break.'" Some but not all children had mastered the use of the iPad to voice frustration.

Leave Me Alone Another reoccurring theme was children using the iPad and Proloquo2Go to voice their desire to be left alone. Perri commented that Cory "gets *so* excited" and "feels so empowered" when he presses the icon that says, "I need my space." She explained that he did not attempt to say the phrase or "even have the concept before" he started using Proloquo2Go. His use of it caused his teachers to view him differently. "That was one of the first things that he was able to say that everyone realized he's purposefully saying this," she told me. "He would find, 'I need my space,' and gently push me away, in class. That was the time that it kicked into the teachers. They were laughing, like, 'Oh my gosh, yeah. He told you he needs his space.'"

For some parents, having their child express resistance in this manner was welcomed because they associated it with "typical" behavior. For example, Vanessa mentioned that "starting late summer, [Moira] got into the 'Get out of my room' thing, which is great, because that's very typical" for a ten-year-old. Nelson wished he could communicate with Stephanie even if she were mad at him. "It's just a typical conversation with a teenager," Nelson said. "Throw a tantrum, and go upstairs, and, 'I'm going to listen to rock music, and I hate you, Dad.'" Marisa offered an alternative interpretation, insisting that Stephanie found other ways to voice resistance. As Marisa described it, "In her way, in her world, she still do. ... Sometimes she go upstairs, put the [air-conditioning on], and go to bed, because she know here is downstairs, and here the AC is freezing." Speaking as Stephanie, Marisa explained her daughter's act of rebellion: "Because I'm going to be in my bed, blankets and everything at night, going to be OK. So, you're going to be proud downstairs."

I experienced a bit of this resistance directly. During my long interview with Nelson and Marisa, Stephanie sat on a stair landing in their apartment, bouncing on a big blue exercise ball while watching videos on an iPad. At multiple points throughout the evening, Stephanie leaned around the corner of the wall to reach the light switch and turned the lights off. Each time, her parents asked her to turn the lights back on, to which she complied after a few seconds. While it was not the same as using embodied oral speech to voice displeasure with her parents (or the researcher in the dining room taking away her parents' undivided attention), "it's like any other child," Marisa explained. Children voiced a need for space not only through words but also through their actions.

I Don't Like Parents valued AAC devices as a tool for children to express their preferences. Donna thought that it was an "extremely important thing" for Sam to say, "I don't like" as well as "I like." "If you don't like stewed prunes, then I'm not going to make you eat stewed prunes anymore," she offered as a particularly salient example. "I don't really give him those, but the 'I don't like' is just, God, almost more important than the 'I like.'" Donna also described how Sam had used his AAC device to say that he didn't like being excluded from a conversation transpiring in front of him. Donna recounted, "Me and my mom one day were talking, and all of a sudden he's like, 'Ignore.' We're like, 'Where did that come from?' Somehow he found in his device 'Ignore,' because we were ignoring him. He wanted to be included in the conversation." Nonspeaking children contend with various challenges in order to give voice to their own needs for safety, relief from pain, and urges to resist. The challenges are partly due to their speech impairments, but also due to technological and environmental shortcomings.

Surveillance and Countersurveillance

While voice calling (i.e., telephone, FaceTime Audio) is a mediated channel in which no record of communication content is usually stored by default (unlike metadata such as the nearest cell tower or length of time spent on the call), speaking through an AAC device isn't necessarily as ephemeral. One feature of Proloquo2Go that prolongs the accessibility of social information is the app's Recents View. AssistiveWare explains in the Proloquo2Go manual that the Recents View toolbar "gives quick access to any sentence that was pressed in the Message Window or spoken in the Typing View some time during the last week."[53] One can tap a message listed in the Recents View in order to speak it. The designers of Proloquo2Go define "recent" in increments of the last fifteen minutes, last hour, earlier today, yesterday, and day before yesterday. AssistiveWare describes the feature as helpful for users because it "can speed up the communication process, as you can easily speak these messages again."[54] The stated purpose of the feature is ease of navigating and recalling the stored data that a user creates on their personal computer, akin to the recent website history bar on an Internet browser.

Implicit above is that the person primarily handling the mobile device and app is the nonspeaking individual. This is not always the case. Having complex communication needs creates its own set of conditions that blurs boundaries regarding voice and privacy for children under age thirteen.[55] For example, Vanessa said that Moira "goes back to her recents" and "she'll look around in there." Sometimes Vanessa will also take a look at Recents View "when I want to see what [Moira] did at school." The log is not only explicitly a persistent record of speech but also implicitly a record of action. Vanessa cannot *see* what Moira did at school (say, through video recordings), but she can surmise as much based on what she said. Vanessa's interest in monitoring Moira's communication stems from legitimate concerns for Moira's safety and well-being, as noted earlier.

Besides parents, clinicians too have a vested interest in recording and analyzing language samples. Starting in the early 2000s, assistive technology companies such as Prentke Romich began to install automated language activity monitor systems in their AAC devices. These systems not only recorded content but time stamps as well. As the field of speech-language pathology began to see itself more as science than as art, data collection guiding patient evaluation also shifted from qualitative to more quantitative methods.[56] From the outset of electronic AAC device development, privacy safeguards were a central concern, both for those communicating through an AAC system and their conversation partners.[57] One experienced AAC specialist with whom I spoke recalled that automated language activity monitoring "came up as a controversy a couple years ago in the field," and "a lot of adult AACers viewed that as a violation of their privacy." She thought that "the intentions behind data trackers are good, but the controversy that it stirred up was really interesting."

If mainstream iPads are more "user friendly" than traditional dedicated AAC devices, then so are the means to access data stored on the machine. One consequence of this shift in usability is that a number of parents discussed Recents View as a tool for countersurveillance on schools' noncompliance with the legally binding education plan outlined in their child's IEP. For instance, Issac's mom, Sara, said, "I think it's good to know that you can go through and make sure that it's really being used" at school. She could hold Issac's teachers and therapists accountable for facilitating as well as encouraging that use.

But in my interview with Karun, she navigated to the Recents View page on Proloquo2Go on Pargev's school iPad in order to voice her frustration with Pargev's teachers and therapists. "When you go to the Recents, I don't see anything," she pointed out. When she looked up what Pargev had spoken the day before, the only word listed was "after." "They're not doing enough," she complained. "Now you see. There is nothing." As we navigated the Recents View feature together, Karun also made sure not to alter anything in order to use it as evidence. "Maybe I should meet with the teacher and ask her. ... I don't want to clear it now. I just want to ask the teacher what's going on."

While the AAC specialist above was aware of the machine tracking utterances, other clinicians and teachers may lack awareness of both the feature and that some parents are watching and interpreting the presence or absence of logged data. The absence of voice through non-use was inherently meaningful to Sara and Karun, and this determination was only possible through surveillance and countersurveillance measures.

A Partial Promise of Voice Output Communication Aids

The parent stories above provide a unique case for interrogating the socially constructed meanings of technologized voice and its partial promise for recognition and response. Contrary to press reports, mere access to mobile media devices with AAC apps does not ensure the "full" expression of voice. What this sense of vocal fullness means also differs across families depending on their access to embodied, objectified, and institutional cultural capital. The potential for these technological systems to enable voice for nonspeaking and minimally speaking individuals will only be partially realized while various concerns are not heard.

As long as the media treats voice as a singular and individual entity, issues regarding who in the household is responsible for promoting and protecting nonspeaking children's voices go unaddressed. The promise of AAC devices to support children's voice is inseparable from the labor of others who upkeep, administer, and moderate the technology on a daily basis, which largely tends to be already-busy mothers.[58]

As long as dominant discourse frames voice as a static entity—an object to find, give, or unlock—then fluctuations in children's development and their environment will not adequately be taken into account. For instance, children may or may not produce more speech by virtue of using the iPad and Proloquo2Go. The software may or may not fully address their vocal needs as they physically change and grow into adolescence. Voices evolve not only biologically but also culturally and socially.

As long as synthetic speech is primarily reserved for Western, English-speaking white people, then inequality among nonspeaking individuals from other backgrounds will be reproduced. Each voice, literary philosopher Roland Barthes writes, has a particular "grain," or traces of the body.[59] The traces of certain bodies, though, are rendered invisible through synthetic speech.[60] Engineers, with limited conceptions of what it means to talk and what "natural" speech sounds like for an "average" person, whisper beneath every inflection of synthetic voice.[61]

As long as professionals are the ones primarily defining the terminology around voice and AAC, then nonspeaking individuals will have less of a say. Therapists and clinicians need to be more sensitive to how families from different cultural, class, and educational backgrounds understand as well as name these technologies, as it can differ from their own training or personal beliefs. Kameelah, for example, told of a more open-ended, elastic approach to referencing her son's iPad with Proloquo2Go, searching for "the perfect word that also leaves it open" to fluctuations in her son, technology, and society. In a sense, Kameelah was queering voice by resisting discrete categories and describing an ongoing, uncertain act of transformation.

As long as AAC is seen as the primary tool of voice production for nonspeaking individuals, then there will be missed opportunities to support more creative expressions of voice through other media.[62] Parents talked about their children drawing on a larger ecology of speech tools, including interactive games and apps. While the children in the study were ages three to thirteen, and not yet using platforms such Facebook, Instagram, and Snapchat, social media (particularly those with a focus on visual communication) offer another key space for nonspeaking teenagers and adults to express themselves.[63]

And lastly, as long as institutions are unprepared to truly hear nonspeaking youth and adults or interrogate the history of listening publics and meanings of listening to disabled publics, the promise of the iPad and Proloquo2Go as communication aids will be unfulfilled.[64] Moira, for instance, had access to a mode of speech production in theory through her mobile device, but when she left home without it, her local police force was wholly unprepared to handle her sudden disappearance from home and reappearance at Starbucks.

Conclusion

Parents across socioeconomic class construct their own unique meanings of voice—at times embodied, performative, and political—and do so in relation to their nonspeaking child, affordances of the speech technology, and often-discriminatory practices of the speaking world. While this chapter explored how nonspeaking children's agency is negotiated across technologies and bodies, the following chapters focus on clashes of agency between parents and schools. I'll start this discussion from the outside—literally, with the cases that enclose tablet computers—and work my way in.

3 Making a Case for iPad Cases: What Is a Mobile Communication Device?

Meryl: What are your thoughts on Danny's iPad case?

Luke's dad, Rob: It's a mixed bag.

In *The Sociological Imagination*, C. Wright Mills charged those who study the social world to illuminate the often-imperceptible links between individuals' intimate stories and narratives told within the wider society.[1] Sociologists and anthropologists refer to this process as "making the familiar strange."[2] Social researchers of communication and technology focus in particular on how individual and collective media practices reflect as well as create societal tensions, as in private and public, control and resistance, structure and agency.[3] Taking on Mills's challenge, this chapter explores how systems of power and networks of social relations not only shape the material aspects of information and communication technologies but also the socially constructed meanings generated through media use.[4] I do so by concentrating on an overlooked component of our personal mobile devices: their cases.

Thriving global and local industries have developed around mobile accessories and peripherals, such as selfie sticks, iPhone chargers, and Bluetooth-enabled speakers.[5] Cases, though ubiquitous among the mobile media-using populace, remain largely invisible to communication researchers. The empirical research on mobile accessories that does exist focuses more on phones than tablets, and suggests that accessories primarily function as status markers and personal fashion statements.[6]

These are, however, only a few ways of understanding mobile device cases. In this chapter, I make the case for making cases strange. I was initially surprised by how much parents had to say about iPad cases, both unprompted during my initial home observations and when asked directly about them in follow-up interviews. In fact, it became difficult to fully understand what the iPad meant to parents without *also* asking them to share what they thought about the iPad's case. As Luke's dad, Rob, notes at the top of this chapter, an iPad case is also "a mixed bag"—it evokes multilayered and sometimes-conflicting meanings. These ideas inform what people do—and do not do—with mobile communication technologies in accordance with their values, beliefs, and ideals.

Some of the families I spent time with bought their iPads and iPad cases using earned income or grants from nonprofit agencies. The school or school district also paid in certain instances. Sometimes, one party (e.g., parents) owned the iPad and the other (e.g., the school) owned the case. Both the technologies and technological accessories that ended up in children's hands were directly as well as indirectly shaped by the broader political, technical, and sociocultural systems at play.

I begin this chapter with a brief overview of how existing scholarship provides theoretical and conceptual tools for situating mobile accessories within communication studies. Then I describe three main themes regarding cases that emerged through my interviews and observations: cases as political, technical, and sociocultural objects. Third, I share three stories that illustrate how these themes intersect and overlap in families' lives—all shaped by parents' varied access to economic, social, and cultural capital. Lastly, I discuss broader implications for how researchers might more fully understand meaning and materiality in mobile communication by getting a better handle on cases.

Theorizing Mobile Accessories and Cases

People can form emotional attachments to their personal communication devices—objects that store within them a biographical "inner history."[7] In a literal sense, smartphone and tablet computer cases enable individuals to *hold on to* their costly mobile technologies more firmly and assuredly as they contend with the unexpected (e.g., dropped devices and cracked screens). Figuratively, people *take hold of* the world, and the world takes hold of them, through the connectivity and communication afforded by graspable devices. Mobile cases are both part of and apart from technologies that are "always-on/always-on-you."[8]

From a symbolic interactionist perspective, cases have social meanings.[9] Technologies are more than functional; they symbolically play communicative roles in our lives.[10] Cultural anthropologists of media practices contend that the tools we use to say things about the world say things to the world about us too.[11] Mobile devices and their accessories display status and can outwardly express an individual's sense of self.[12] For instance, a fellow communication scholar, who studies Internet culture, has a cell phone case covered in a pattern of her favorite emoji.

"Cases" and "accessories" also have a history as technological terms. Beginning with the Apple II, the main components of a personal computer have generally had some sort of casing, an enclosure usually made of steel, aluminum, or plastic. Cases are accessories to mobile devices, but mobiles can also be accessories to other computers. Personal digital assistants, for example, were marketed in the mid-1990s as accessories to the desktops and laptops that people already owned.[13] Palm Computing's late 1990s' marketing campaigns also imagined "women as adornments and assistants to men in the workforce and as primary (often sole) familial workers in the domestic sphere."[14] Mobile devices have their own accessories, are accessories (to a person's body or their personal computer), and the discourses around mobile devices can position their bearers as accessories too.

Personal digital assistants and laptops were also the first mass-market portable computers to be used for AAC, despite the rhetoric around iPads emphasizing their "revolutionary" nature as assistive speech technologies.[15] Yet while those earlier portable computers were at first marketed broadly as being primarily suited for work, Apple introduced the iPad to the consumer market as a computer designed expressly for leisure, recreation, and creativity.[16] This initial stylistic framing of the iPad contrasts with the manner in which assistive technologies are generally characterized, highlighting medical necessity and task completion over self-expression and identity play.[17]

How Cases Matter

The overview above—situated in media sociology, cultural anthropology, and critical studies—provides a starting point for understanding the meanings that people derive from their mobile cases as well as from their cased (or uncased) mobile devices. The following section details how mobile device cases matter to parents of iPad AAC users as material and symbolic artifacts, mapped along various political, technical, and sociocultural dimensions.

Political

When bundled with educational technologies, cases cannot be understood outside institutional politics. In this respect, cases represent *parental proactivity* outside the limitations set by schools and school districts, and a means for the child to *prove their competence* in light of staff and administrators presuming the child's incompetence in handling the delicate iPad.

Parental Proactivity Researching, buying, or making an iPad case was one simple way that some parents felt they could help their child when otherwise tied up by school district red tape. Many parents reported that the assistive technology assessment and delivery process across various school districts in Southern California could take years as well as involve many false starts. "About three years we start to fighting for communication device," recounted Stephanie's mom, Marisa.

Instead of waiting for schools or insurance companies, a number of families across income levels decided to begin by independently obtaining an iPad and iPad case for their child. In this consumer-driven model of assistive technology adoption, the client's platform preference guides the purchase, in contrast to a clinical model guided by a professional's judgment of the client's needs.[18] "My wife and I have always tried to be proactive, and tried to go one step ahead as much as we can for our child," remarked Stephanie's dad, Nelson—especially when the school district was one step, if not more, behind. Although Nelson and Marisa indicated to me that their average yearly combined household income was between $25,000 and $49,000 (slightly below middle class in Southern California), "I went and got my own [iPad]," said Nelson, "just so that whether or not they give us one, we're going to try to help her with it."

Finding the right iPad case was particularly important to Nelson, a self-described computer enthusiast. "On my own, I went out and got that Griffin [case], which is actually a little bit better than the OtterBox [case]," he detailed, referring to two mobile device case brands known for their toughness and durability. "I've done my homework," he declared, "and it's supposed to be the military version." At the IEP meeting for Stephanie in which he and his wife first broached the topic of their daughter using an iPad for AAC, "[the district] had a technology person there that was like, 'You guys know more about this stuff than we do,'" Nelson told me proudly.

When the school district ultimately did provide Stephanie with an iPad for communication, the case that the district selected (without Nelson or his wife's input) was far flimsier than the Griffin. When Rachel suggested that the family purchase a new case for the home iPad to differentiate it from the school iPad (as both were colored black), Nelson replied defensively, "Maybe the school one, because that's a Griffin." Nelson's research into and purchase of the Griffin case exemplifies how iPad cases can symbolize parental proactivity by turning to the private market when social services fall far short of expectations.

Proving Competence In the hands of people with or without disabilities, it does not take much for an iPad to break. Thomas, for instance, had been using an iPad that the regional center had loaned his family until the school provided one for him. Thomas's mom, Daisy, thought that it was cruelly ironic that Thomas shattered the regional center's iPad because "the week before, Caren was saying that she'd recommend a nice case [for the school-issued iPad], the heavy duty." When I complemented Danny's parents on their cool-looking iPad case, Danny's dad, Peter, replied quickly, "That one has been thrown across the garage" and yet managed to withstand the impact. Concerns about the iPad breaking along with the resultant replacement costs are certainly not unwarranted.

If the child broke the school-issued device while on school grounds, then the school or school district (depending on the funding source) was generally responsible for paying for a replacement iPad. Noted one AAC specialist I spoke with, "The good thing, even if the screen breaks or anything like that, is we do have a loan inventory [in the school district]. If the system is written into a student's IEP, we as a district have to legally provide it." If the child broke the iPad or lost it outside school, it was unclear if the school district or parents would ultimately be held financially responsible.

Some parents felt like school districts took this concern too far, leading to a default characterization of their child as inherently reckless with technology. Schools frequently operate under the assumption that children with disabilities must prove their competence in order to be granted it.[19] Less privileged parents had fewer resources at their disposal and greater difficulty in defending their child from this characterization. Nelson commented, "[The school district was] very hesitant giving us this device in particular because they thought [Stephanie] was going to drop [the iPad], or she wouldn't be able to hold on to it." Nelson did not agree with the school district's depiction of his daughter. With the iPad that Nelson and his wife purchased on their own, he said that "[Stephanie has] dropped it a couple of times, but as you can see, the way [the case is] designed, it's really easy for her to take it, and walk around with it, and set it down." A poorly protected iPad might lead to breakage, putting Stephanie in a vulnerable position to be deemed incompetent by the school district, giving it leverage to deny her services.

In another school district, though, the AAC coordinator I spoke with downplayed parental responsibility for replacing a broken iPad. She explained, "Basically, when I sit down with parents and have them sign that agreement, I really emphasize that the agreement is not a legally binding contract. If the device breaks and it's not a purposeful accident, we're not going to come after them. It's just more of an acknowledgment that the equipment is very valuable, and they are going to do their best to keep it in a safe place and not put it in a situation where it's going to become damaged." Seeing as this was not the message relayed to a number of the families I spoke with, these policies seem to vary widely across school districts, at least in terms of enforcement.

The concern over children breaking the iPad also shaped some district assistive technology policies even after iPads were distributed. In Moira's old school district, her mother, Vanessa, remarked that out of concern "that kids were throwing them in the pool or breaking them," the school district suddenly "changed their policy and said that iPads only remained on campus." This was after Moira had already been using the iPad and Proloquo2Go for over a year. Vanessa believed that this policy change was an abuse of power and breach of Moira's IEP, which explicitly stated that her AAC iPad was for community use as well as for use at home as part of Moira's autism services, which also happened to be funded through the school district. Vanessa told me, "I wrote [the school district] and said, 'This is in violation of her IEP. I am asking that you give me a window of opportunity to purchase her a device for the home.'"

But one day, Moira's iPad went to school and never came home. "They totally just disregarded my request that they give me at least thirty days to purchase her a device," recounted Vanessa. She did not have other resources at her disposal for advocacy, such as the finances to hire a lawyer or a social network of people to recommend further recourse. The school district's preemptive measure was shortsighted, more focused on protecting the iPad in the short term than promoting Moira in the long run.

Moira and Vanessa eventually moved to a new town, where much to Vanessa's relief, Moira's new school district provided her with an iPad and case that Moira was allowed to take wherever she went. Vanessa was "bothered [that the old district] made this blanket decision" because Moira "really takes good care of [the iPad] for a kid." As proof, she stood up during our interview, went to Moira's bedroom, and asked Moira to bring her iPad into the living room to show me. "This case is great," remarked Vanessa, holding it up. "[Her new school district] provided this one." The new protective case was chunky and pink, made of rigid foam.

The straps, however, had fallen off, leaving Moira to only use the case's built-in handle. Vanessa did not have the disposable income to purchase a new case on her own, yet the family made do with it. "She can't just be hands free with it, but she doesn't mind carrying it," remarked Vanessa. She stressed Moira's responsibility with the iPad, saying, "[Moira] has *total* ownership of it. It's the first thing she grabs before we leave the house." Moira's new iPad case enabled her to demonstrate her competence, whereas her old school district had presumed her incompetence. In this way, cases could both shield the iPad and uphold the child's dignity.

Technical

While tablet computers have particular technical properties such as screen size and weight, iPad cases have their own as well. These properties can materially shape the relationship of the iPad to *space* and *bodies*. iPad cases both afford and constrain technical possibilities, opening or closing up social settings in which to use the technology, and promoting or preventing certain kinds of social interactions. A tension emerged between protecting the iPad from the elements and yet exposing the screen for the child's easy use.

Relationship to Space Compared to the pristine white walls of subdued Apple stores, iPads are used in much dirtier and louder spaces. Interviewing families in the Los Angeles area, the pool and beach came up as potential sites where the child might use the iPad, but only with a waterproof case. More privileged families spoke of owning multiple iPad cases, and that these options afforded them a certain freedom of movement with the device.

For instance, Eric's mom, Anne, was excited that Eric would be going to a swim camp for children with disabilities during the upcoming summer. When I asked her if Eric would be carrying his iPad around camp, she replied, "I think we'll probably have to get a better waterproof cover for it so it isn't affected." Danny's dad, Peter, mentioned, "We do not take it to the pool, but I have taken it to the beach because I have a case that's waterproof. I just switch it over and let [Danny] use my waterproof [one]."

Families need not go as far as the beach to find a context in which the iPad requires protection, seeing as water and sand are not the only substances that might harm an unprotected tablet or phone. Nelson noted, "I remember when [Stephanie] took her [Mom's] smartphone and dipped it in ranch dressing accidentally. She likes to dip stuff, and it survived. We just cleaned it." As any parent knows, domestic spaces such as kitchens and bathrooms introduce elements that compromise the integrity of any object or item than cannot be washed or wiped down. A rubber case might not protect an iPhone from getting dunked in the toilet, but it would at least shield the device from condiments and crumbs.

Besides expanding the geography of communication, a case can widen the proximity of audible speech, specifically in loud spaces. Special iPad cases now come with speakers that amplify sound output from the iPad. Moira's mom, Vanessa, complained about the iPad's limited volume. "So if we're at a restaurant or in a loud place," she said, "then people can't hear [Moira]." Instead, "what I started doing recently," explained Vanessa, "and I don't know if this comes off as *rude*, but if they can't hear her, then I pick up the iPad and I show them [on-screen] what it says." Vanessa was curious, since "now they make cases with additional speakers," if such a case might help Moira speak in those situations. An iPad case with built-in speakers could enable a person to communicate more independently, boosting the volume of their voice output and giving them greater freedom to converse with colocated others in loud spaces.

Relationship to Bodies Some iPad cases made it harder or easier for a person to use the device while their body was in a certain position. A number came with built-in or additional stands to angle the iPad screen toward a person while seated at a table. Cory's mom, Perri, was slightly dissatisfied with Cory's iPad case, which had a handle that could also be used as a built-in stand. "I was hoping that this," Perri said, while sitting next to me at a café table as she gripped the handle, "would be a little wider so that when it stood up like this"—she rested the handle on the table and bent the iPad backward—"[Cory] wouldn't knock it over, or when it was up like this"—she flipped the iPad and demonstrated how the handle raises one end of the iPad up about thirty degrees off the table—"it has a better angle." In the interim, Perri bought another stand for Cory's iPad case "that puts it up more to a forty-five-degree angle, ... but it's like a big metal thing, so. ..." She found a temporary but unsatisfying fix so that Cory could use the iPad while seated at a table.

The case material could similarly promote or prevent easy handling. River's dad, Mark, recalled, "When I got his [iPad] Mini, I got a really light case, but it had almost like a velvet. ... It was almost sticky. Even though it was smooth, you get a little grip on it." Hands fumble with complicated cases, and Mark noted that the same case also had a design flaw. "It didn't take too long before that one got just destroyed because the cover just flapped," he recounted. "It didn't have like a little Velcro or anything, so he was grabbing it by that."

This coverage issue also poses a problem when an AAC user wants to communicate quickly through the technology. Talen's mom, Kameelah, mentioned she was "looking for carriers the other day, and I didn't find one. … I was like, where is one that the face will be out? Like, it'll be protected, he can carry it, but he will just lift it up and poke. Instead of lift it up, open it up, pull it out, and do all of that." The new case that Mark's brother, "the Mac guy" of his family, had gotten River for Christmas sounded closer to Kameelah's ideal case. Said Mark, "It's mostly nylon strap type of material. It's got a little thing you could slide your hand in, on the back. You can flip it like a little candy vendor tray." Despite their sleek appearance, iPads do not seamlessly meld to bodies; iPad cases expose those seams.

Sociocultural

Social Stigma and Acceptance Many parents mentioned that while their children were socially ostracized in multiple ways, they were more easily accepted among typically developing peers due to the iPad being perceived as something "cool." Danny's mom, Alice, observed that iPads "tend to be a little more socially acceptable [than a dedicated AAC device]. A person sees someone like Danny with an iPad, and it's not weird." Appealing cases also carry a certain degree of objectified cultural capital among other middle- and upper-class parents. Alice proudly noted, "I've had lots of parents comment on, 'Where did you get that case? I'd love one for my kid.' It never even occurs to them that my kid has a disability. It's just, 'Wow! That's a really cool case.'" As highly visible markers, cases enable a way to blend in.

Gendered Labor While the iPad case may partially protect the iPad, the technology does not take care of itself. Cases can make life easier in some ways, but they can end up creating additional material labor and "emotion work" for caretakers overseeing the child and their technology use at home—frequently expected to be mothers.[20] For instance, Kevin's mom, Rebecca, said that even though Kevin's iPad case was a "rubberized heavy-duty" one, she had to take care to put it in his backpack each night just right. "It doesn't have a cover on the front, so it just goes in his backpack. But I just have to put it so that it's not … [so] the screen isn't exposed to the outside of the backpack."

Thomas's mom, Daisy, recounted that there had been a number of occasions when Thomas's iPad had been misplaced at school because of its case. She turned the iPad over and pointed to a place on the back of the case where a label with Thomas's name had once been. Another boy in Thomas's class also uses an iPad for AAC with the exact same case as Thomas—a case made of army-green-colored hard plastic. Daisy said that Thomas and his classmate have often accidentally ended up with one another's iPads at the end of the day, only to be discovered by their moms once the boys have taken the school bus home.

When these accidents happen, Daisy explained that Thomas "gets frustrated" because he can't communicate, and because of this, he acts out. Although the other boy also uses Proloquo2Go, the software is customized in such a way that would make it difficult for another person to borrow and use. Thomas additionally uses part of Proloquo2Go as a visual schedule of his daily activities, which helps him with self-regulation and feeling calm. To add to the families' frustrations, one of the moms then has to drive over to the others' house to swap iPads. Confusion with the cases at school produces extra labor, materially and emotionally, for Daisy and the other boy's mother.

Case Studies

The noun "case" has other meanings besides a physical object. The case, as a concept, is at the heart of qualitative social science. Charles Ragin and Howard Becker contend that researchers should ask what any given social phenomenon is "a case of" in their analyses.[21] The case is where ideas and evidence interact. In the examples above, I have provided cases, per Ragin and Becker's definition, of parents making sense of their child's technology use, partly informed by their understandings of the technology's accessories. Cases play a role in how parents comprehend the affordances and constraints not only of the iPad itself but also what they imagine can and cannot be done considering the social, cultural, and political contexts that undergird potential uses of the device. The following extended examples demonstrate how the material and symbolic dimensions of cases overlap in parents' meaning-making processes.

Beatriz

Beatriz lives in a one-bedroom apartment with her parents, David and Pilar, and her older sister, Ariana. Beatriz and Ariana share the bedroom, while David and Pilar sleep in the living room. When I visited Beatriz and her family, we sat spread out across her parents' bed, a small sofa, and the rolling chairs for two desktop computers. David and Pilar are immigrants from Mexico; he is fluent in English, and she is not, but David graciously translated for me.

Beatriz initially had a thick rubber, bubble-gum-pink-colored Big Grips case on her iPad. When I first interviewed David in 2013, he told me the case was "helping a lot because several times she has dropped the iPad, and it just bounces. It never breaks!" He also liked this brand case "because it was thick but it was easy to grab." School had selected the case, and while David was happy with what they chose, the process was not transparent to him. "I try to find more," he said of the Big Grips case, "but I cannot find. I don't know where they bought it." When I interviewed David and Pilar a year later, I noticed that the pink Big Grips case was gone, replaced with a black nylon iPad case. Again, the district's decision was opaque. No one made David, Pilar, or Beatriz aware that they would be replacing the case. "I don't know why they change it," he told me.

Functionally, the new case actually made communication even more difficult for Beatriz. David demonstrated how once the cover on the new case was lifted, revealing the screen, there was no Velcro or magnet to keep it open. The case, it seemed, was designed to stay closed. "For the protection of the device, it's nice," he remarked, "but for daily use, I don't think [so]." David believed that the design of the case impacted how Beatriz used the iPad for communication. "Probably, this has to do with encouragement to use it or not use it, the way [the case is] designed," he said. While the case secured the screen, it constrained communication. To David, the case symbolized the school district's skewed values and priorities. Of the district, he observed astutely, "They put more importance on the protection of the device than on the use of it."

While the school put little thought into how this change might impact Beatriz, David thought that the switch had a far-reaching impact. It shifted the way in which Beatriz interacted with the iPad Mini that he had bought for her to use at home. Noted David, "She takes [the iPad Mini case] off. Maybe that's what she needs all the time. I think she feels she's a little restricted with this one [the school-issued iPad], and that's why on her own [iPad], she wants it to be free." David characterized Beatriz's removal of the iPad Mini case as an act of liberation. She could enact this particular form of resistance without consequences from her school since her father owned the iPad Mini and its case.

Beyond function, the new black iPad case sent a message. It communicated something about Beatriz, but the wrong thing. To David, the new case said, "She's careless, beware now, she's going to drop it, she's going to break it." The new black case was also more stigmatizing than the old pink one. David remarked, "It does look for someone that is careless. It's good for that kind of people, but she takes care of things. She gets it dirty, but she takes care of it by not dropping it, by not throwing it." Just as with Stephanie and Moira, the school district presumed Beatriz's incompetence, symbolized by a case that "doesn't look appealing. It doesn't look chic."

The irony was that Beatriz actually cared a great deal about the aesthetics of mobile device covers. As we were discussing the new black iPad case, Beatriz went into her bedroom and brought back a clear, hard plastic pencil case full of rubber iPod Touch and iPhone cases. "Look!" she said orally as she opened the case to show her mom. David turned to me and explained, "She has this tendency of collecting. She has an obsession for covers." He then turned to Beatriz and asked her, "Right? You like covers?" "Look, look! One ... two ... three!" exclaimed Beatriz, as she handed the cases off into her mother's hand, one by one. Cases are a low-cost toy for Beatriz, David said. "We get them, but we try to get the cheapest." I asked him if she ever put them on the family's mobile devices. "She has, probably instead of playing dolls, she plays dress-up iPad," he pondered. How unfortunate, then, that Beatriz's communication aid was clothed in the equivalent of a drab school uniform.

Luke

Luke's parents bought his iPad to use for AAC, but the school district chose and purchased the iPad case. His mom, Debra, reported, "We wanted to get the ball rolling, so we said, 'Let's just get the iPad and get it going.' Then [the school district] said, 'Oh you need a case, you need a case!'" The school district had various "criterion" for the case they would buy, explained Luke's dad, Rob—rules that he did not fully agree with. One was a case with a speaker because "they wanted more volume out of it, which I thought was silly, because I think [the iPad's] plenty loud." The school also wanted a case that could be carried. Rob thought that "the case that came with it ... was reasonably carryable." But the school ended up choosing the iAdapter, a case with a built-in speaker and carrying weight, combined with the iPad, of 2.5 pounds. According to the installation and operating instructions for the iAdapter, there are nine screws that hold the front and back shells of the case together, making the iAdapter a semipermanent part of the iPad.[22]

While amplification and portability were the school's top priorities, those things were "relatively arbitrary" to Rob. Luke's parents valued the iAdapter's protection and sturdiness. "It has saved us a couple times when I've dropped it," said Debra. Rob mentioned, "I broke [Debra's] iPad. It wouldn't have happened if I'd had that case." While the school wanted the iAdapter because the case afforded mobility, Rob liked that the iAdapter case immobilized the iPad. "What's good about [the case]," remarked Rob, "is the stand and all that kind of stuff, and the fact that it's designed to be used at a table." The stand "holds it really firmly, so you can hammer away on the thing like it was a built-in screen. For classroom instruction, for sitting around and looking at things, that's really good." The iAdapter case temporarily transformed the iPad into a desktop computer.

Compared to Rob, though, Debra had much more of a love-hate relationship with the iAdapter case. To her, the large, heavy shell took away from the aesthetics of the iPad. "I think it looks ridiculous. ... It destroys the fact—I think, we think—of how sleek an iPad is," she protested. During my discussion about the case with Debra, Luke's teenage brother, Spencer, walked through the kitchen and chimed in, "I don't like the look of it. I can yell all about it and trash talk it, but it will still be there." "It's horrible. We make fun of it all the time," said Debra. "It looks like an old Dynavox." Debra felt like the iAdapter case made the iPad look like a dedicated AAC device, and perceived this as a negative and socially stigmatizing association.

When Rob came home and joined in on the interview, he expressed a different view from his wife. Rob (who was unaware of Debra's earlier comments) remarked, "What people don't like about it is the, 'Hey, it makes it look like one of *those* devices.' I agree, it does that. ... Some people are really sensitive to those issues. I don't. We're not." Instead, Rob expressed that the stigmatization of dedicated devices was not "as big a deal" as the social stigma surrounding autistic individuals stimming in public. Said Rob, "[Luke] running around and jumping in circles and doing other things are what identify him as a little bit odd!" While the case on the iPad could be changed, Luke's stimming was an immutable part of his identity.

Pargev

As I noted in chapter 1, Pargev and his family had recently fled from war-torn Syria and sought asylum in the United States. Neither of his parents were employed at the time of my observations and interviews with the family. Pargev's dad, a practicing doctor in Syria, was looking for work in the United States. Karun told me she was "not working, to take care of [Pargev]," but "I'm starting a new business with scarves and jewelry." She showed me the room where she crafted her handmade goods—the space in their two-bedroom apartment likely designated by the building's developers as the home office. At the end of our interview, around the holiday season, she handed me a business card (also self-produced), saying that I should come to her if I wanted to purchase any gifts.

At my first observation in August 2013, Karun expressed concern that the iPad the school had provided for Pargev was too heavy, and its weight deterred him from carrying it places and communicating with it outside the home. She asked Rachel, the speech-language pathologist, if the school might supply Pargev with an iPad Mini instead of the full one he currently had—a Mini that he could carry in a bag with a strap. Rachel responded that the school could give him the iPad Mini, with the caveat that "the cases that have straps, they're not as durable" as the strapless iPad case that the school originally provided.

During my second observation, two weeks later, Karun told Rachel, "By the way, Naomi [the school district's assistive technology specialist] says she's gonna get the Mini iPad. It will be really nice to take it wherever we go. The size is perfect. She didn't say when. You want to remind her?" "It's probably better if it comes from you," Rachel responded. "OK," replied Karun. "It's been a while." At another point during the session, Karun asked Rachel if she could also talk to the school speech-language pathologist on her behalf about making changes to the words available to Pargev on his Proloquo2Go software. Karun told Rachel, "It's better if you ask because you're more professional." This exchange highlighted the distinctions in institutionalized cultural capital between Karun and Rachel, and Karun's desire for Rachel to advocate as an intermediary on her behalf. Karun thought that the school was more likely to respect and listen to Rachel than herself, toward whom they had been unresponsive.

Considering Karun's crafting initiative and frustration in working with the school, it was not surprising that she expressed interest in making a case for the iPad Mini. At the time of our interview in November, two months after my second observation, the school district had still not provided the Mini. Karun explained her motivation for making a case for Pargev as such: "The best way is if I try harder, when I take him out, to take the iPad with me because it's big, and he is forgetting it. If this was smaller and hangable here somewhere [points to Pargev's chest], as a messenger bag or somewhere. ... I will eventually make [the iPad Mini case] because I got an e-mail from the school district. They will exchange [his current iPad] with Mini." Karun's hands were tied in terms of making any progress with the school district, but a handmade iPad Mini case was one way she could provide for her child.

Grasping the Encased, Unencased, and Reencased

In mobile communication research, device cases and accessories are a largely understudied area. Talking with iPad owners and users about their cases and covers presents an opportunity for examining the interplay between the symbolic and material. How individuals perceive mobile communication technologies requires understanding how their devices are encased, unencased, and reencased. Both in relation to and independent from the devices themselves, cases have their own complex social meanings, at least in part informed by users' class status and their relationships to the social institutions shaping the use of encased devices. Below, I discuss these findings in relation to three strands of research on communication technology use: the notion of technology as materially and symbolically *durable*, the *mobility* of mobile media as a static quality, and the links between *identity* and digital communication technologies.

Durability

Bruno Latour memorably wrote that "technology is society made durable," meaning that technologies reflect some fleeting moment of social stability in which they are initially constructed.[23] Leah Lievrouw extends this formulation from the field of science and technology studies to media studies, describing technology as *communication* made durable.[24] Others have argued that durability is itself an unstable concept—one continually made and unmade through sociotechnical practices.[25]

The ways in which iPad cases are incorporated into families' lives reflects these various conceptions of durability and the friction between them. How durable a communication technology is depends on the economic, political, and social infrastructures in which the technology is entrenched. Apple, like any profit-driven computer company, purposefully designs its hardware and software to be replaced by successive generations of technology, also known as planned obsolescence.[26] In the same way that rare comic book collectors use clear plastic sheets to protect their expensive and delicate objects, cases can prolong the material durability of an iPad depending on the environmental conditions to which it is exposed. As parents mentioned, this includes sand, water, and even salad dressing.

Durability is also relational, for it is partly determined by insurance policies and warranties. Insurance companies do not consider iPads to be "durable medical equipment," and thus the tablets are not covered when requested by parents for their child to use as an AAC device.[27] In this absence, school districts enact the power to assign their own value to durability. They want to use cases to protect their investments in technology over time, sometimes more so than their long-term investment in students. As in the instances of Stephanie, Moira, and Beatriz, those students from lower-class backgrounds are often the most vulnerable when these decisions are not in their best interests. Cases underscore the iPad's inherent material and symbolic fragility as well as the fragile relationship between citizens and social services.

Mobility

Across public, private, and hybrid spaces, mobile device cases allow us to reflect on both how media are housed and how homes are mediated. With straps and handles, cases enable mobility outside the home, as desired by Moira and Pargev's moms. Yet parents (frequently mothers) had to position encased devices in just the right way for transportation (e.g., inside backpacks). Mobile telecommunications technologies in the postwar era tend to reflect desires for portable forms of domesticity. Lynn Spigel identifies an evolution from "theatrical models of home to vehicular models of home," and finds evidence for such a shift in the marketing of portable televisions and receivers in the 1960s.[28] The very design of these technologies emphasized their mobility; whereas television sets once looked like living room furniture, portable televisions were packaged inside cabinets resembling luggage.

Portable computing devices have taken various names and forms since the 1980s: luggables, transportables, portables, totables, uprights, laptops, notebooks, personal digital assistants, tablets, and more recently, phablets (a portmanteau of "phone" and "tablet"). Shifting notions of "mobile computers" along with their assorted cases and enclosures are coconstituted by conceptions of work and leisure as well as the relation of the embodied human subject to domestic space. iPad mounting systems, another kind of computer accessory, can attach tablets to the tubing of more mobile wheelchairs and the frames of less mobile beds with Velcro, clamps, and retractable arms.

Parents described how cases strategically immobilized the iPad for use at a table during mealtimes or on a desk in the classroom. Luke's dad stressed this transformation when he lauded the iAdapter case for allowing Luke, as quoted above, to "hammer away on the thing like it was a built-in screen" without fear of the hammering shattering the glass screen. Cases can impact the extent to which a mobile media and communication device comfortably travels within and outside home and school without making device owners and users anxious about breakage and liability.

Identity

Lastly, mobile communication technologies and accessories are not symbolically seamless extensions of the body, but rather sites where identity and personhood are negotiated.[29] This is particularly true for individuals with disabilities, whose experiences with communication technologies are bound up with the interests of those administering and overseeing the use of these tools. Beatriz, for instance, was assigned an iPad case that did not at all reflect her personality or interests. The school never considered letting her have a choice in her case. This in part impeded her desire to communicate using the technology.

By another name, cases are physical analogues to digital skins. A skin refers to the customization of an avatar's body, associated with online modding and gameplay. Like cases, skins are functionally interchangeable in that swapping one out for another does not change the underlying mechanics of the game. Skins also materially and symbolically color and are colored by their online environments. For example, avatars in virtual worlds generally come with a limited range of preset skin tones and matching body parts.[30] Digital skins can reproduce "real-world" hierarchical social structures.

An AAC user's desire to have an iPad case that reflects their identity, and the potential impact on psychological well-being if they are unhappy with the case's look and feel, is also akin to prosthesis aesthetics, and feelings of satisfaction and dissatisfaction with the cosmetic cover (or cosmesis) fitted over a mechanical limb. As with covering the body, each society and culture has particular norms about concealing or revealing artificial limbs, and a certain set of prosthetic options available.[31] Within this possible range, some amputees may choose to cover their artificial limbs with silicon sleeves that approximate their skin tone, while others don more colorful fabric slip-on covers as changeable accessories, and some choose not to cover the metal or wood at all. One might choose to paint the toenails on their prosthetic foot and change their polish color regularly, and another might keep their toenails bare.

Cases, like skins and sleeves, are a means of identity performance through altering the appearance of personal and intimate technologies. Beatriz's dad viewed her "obsession with cases" as akin to doll play and her removal of the case from the family's iPad Mini as an act of emancipation. Her district-issued iPad case did not fit her style. The degree to which a mobile case is stigmatizing (or normalizing) needs to be considered relative to other forms of social stigma that the bearer of the case might experience. While Luke's mom was concerned with the iAdapter case making the iPad more overtly like a piece of assistive technology, for instance, Luke's dad weighed the social stigma associated with the object against the social stigma associated with Luke's stimming behaviors.

Conclusion

At a time when popular discussion of mobile communication emphasizes its ephemerality (e.g., the Whisper app) and evaporation (e.g., cloud computing), cases matter because they are composed of atomic matter—obviously so when brightly colored and boldly textured rubber covers the smooth silver and rose gold of metallic Apple devices. The research presented in this chapter illustrates the political, technical, and sociocultural dimensions of mobile cases along with the centrality of materiality in the study of media and communication technologies. Cases can be physically detached from mobile devices, but are still very much "part" of them, as much as skin is part of the body. Beyond their use as fashion or status symbols, the protective cases that people put on their mobile devices serve as a site for understanding class distinctions and power struggles. Upper-, middle-, and working-class parents have unequal opportunities to access and mobilize economic, social, and cultural capital in order to advance their children's communication rights with these sets of tools.

Politically, cases can symbolize parental proactivity in response to a breakdown in social services as well as the competence of a child presumed to be incompetent by others. Technically, cases have particular material relationships to bodies in space that shape their perceived functionality. iPad cases can also raise issues of self-presentation, as well as gendered expectations of household labor in taking care of domestic technologies. The cases that cover iPads used as assistive technology for speech are in some ways like the cheap silicone that covers smartphones, in other ways similar to the expensive silicone that covers prosthetic limbs, and in other ways akin to the pixels that digitally cover the bodies of avatars.

The study of ordinary device cases unexpectedly opens up key questions in the study of the social uses of mobile communication concerning durability, mobility, and identity. In cultural studies of media, the material condition is often abstracted, be it by the traditional encoding/decoding model of media audience spectatorship, or a detached focus on functionality and institutions. By approaching the sociotechnical quite literally from the outside in, this chapter reminds us that personal computers are objects that once removed from the box, can be—and inevitably are—additionally personalized.

The next chapter further unwraps the iPad and extends the discussion of the symbolic meanings that parents associated with cases, primarily how different cases were sometimes used by parents and professionals to distinguish between at least two different iPads at home—language that differed among more and less privileged parents.

4 The "Fun iPad" and the "Communication iPad": What Is an iPad For?

> We call [Raul's iPad with Proloquo2Go] his "talker." ... We realized we can't call it "an iPad" because my husband had one already. We didn't want him to confuse the two because, truly, we wanted him to know, "This is your voice," and that's it. Then we had a blue cover around it. We made it clear: "This is yours." We put his name on it, and my husband's is the "fun iPad." It has the games, the apps, the cartoons, the music.
>
> —Nina, Raul's mother

> [My] son has the original iPad. That's his "play iPad." ... When he turns this one on, ... he doesn't try and find anything else because now he knows this as his "communication device." Since this case has been on it, it has been nothing but "the communication device."
>
> —Perri, Cory's mother

In the last chapter, I discussed various ways in which parents perceived iPad cases to be more than fashion statements and status symbols, as mobile accessories are usually understood, but as having a wide array of meanings tied to larger political, technical, and social systems. Beyond those core themes, there was another way in which some parents talked about iPad cases and encased iPads. In the quotes above, Nina and Perri describe cases as visual delineators between two types of iPads used by their children at home: first, iPads for *fun* and *play*, and second, iPads for *vocalizing* and *communicating*. "We made it clear" that the two machines were different, Nina said—a distinction visible from the blue cover on the "talker." For Perri, the case cemented the iPad with the Proloquo2Go app as "nothing but 'the communication device.'"

While there has been some research exploring the attitudes of parents and professionals toward the integration of the iPad as an AAC device into education and therapeutic interventions, this research is limited in various ways.[1] It primarily centers on caregivers and therapists of autistic populations, and not other developmental disabilities that can manifest in complex communication needs. It does not report on the income, race, ethnicity, or immigration status of parents. Data are also rarely collected on parental and professional views of the types of mobile applications used by children at home, and if these are seen as being for education and/or entertainment.[2]

This last limitation is particularly important because today's mobile media are capable of being the one tool on which we rely the most. Mobile devices are convergent whether or not we desire them to be; for example, it is difficult to purchase a cell phone nowadays that only sends and receives phone calls.[3] At a time when there are few things that a mobile technology cannot do with the right app installed, I was surprised to hear numerous parents talk about their children ideally using two different iPads, each for limited purposes: fun *or* communication. I was also taken aback that most children in my study (fifteen out of twenty), across socioeconomic groups, had access to two or more iPads when they were outside school. At least one of these iPads had Proloquo2Go installed (owned by the school district or family), and the family owned at least one other iPad (which may or may not have had Proloquo2Go as well).

This chapter explores the following questions: What characterized parents who made distinctions between the "fun iPad" and the "communication iPad"? What language did other parents develop for the two devices? What purposes did these differentiations serve, and for whose benefit? And how does this phenomenon reflect, and perhaps refract, broader understandings of how children's media and technology use is regulated in public and private spaces?[4]

I begin this chapter by examining how more and less privileged families came to use multiple iPads at home. Next, I discuss key differences in how both groups of parents distinguished between devices. I then detail how children's use of iPad apps for fun and communication at home were promoted as well as prevented through "locking" children into or out of the technology—a term that took on multiple meanings. In all, I argue that these various naming practices are not arbitrary; they reflect parents' differential access to economic, social, and cultural capital, and reinforce structural biases.

Demographics of iPad Ownership

One AAC consultant working in a Los Angeles school with a large immigrant population from Mexico relayed to me that a parent had just asked her if Proloquo2Go could be installed on an Android device. She replied, "[No], it's only iOS, only Apple based. That's kind of an impediment for some families, especially lower-income families that can't afford it, unfortunately." While this may be true, the majority of children in my study (75 percent) had access to two or more iPads outside school, at least one of which had the app Proloquo2Go installed.

A 2014 survey of US family media use conducted by the Joan Ganz Cooney Center at Sesame Workshop reports that more than half (55 percent) of two- to ten-year-olds live in households that have at least one tablet device at home.[5] Of the fifteen children in the study with access to two iPads at home, nine were using iPads for AAC that were provided by the school district. The other six children were using iPads with Proloquo2Go that were owned by their parents but brought to school daily.

Ownership of an additional iPad was clustered around the highest-earning families. Of the fifteen with two iPads at home, about half (seven families) reported earning $100,000 a year or more, and the rest were middle- and lower-income (two reported a household income of $50,000 to $99,000 a year, four reported a yearly income of $25,000 to $49,000, and one family reported earning less than $25,000 annually).[6] This distribution is in line with research indicating that tablet ownership among families (specifically with young children) differs significantly across income levels. While over three-quarters (77 percent) of high-income children (greater than $100,000 annually) live in households with a tablet computer, only about one-quarter (27 percent) of low-income children (less than $25,000 a year) have access to one at home.[7]

These numbers only tell part of the story, though. Families with different socioeconomic backgrounds ended up with additional iPads in various ways. Less privileged parents tended to obtain them through outside funding sources. For example, while Talen's parents technically owned both of his iPads, one was a gift from one of Talen's former therapists, and the other was provided through a state grant. For some less privileged families, investment in an iPad required saving up over an extended period of time. Beatriz's dad remarked, half-jokingly, that if "she's gonna be using it, it's worth it to save some money, to stop buying a six-pack [of beer] every weekend."

Moira only ended up with an additional iPad with Proloquo2Go after the district barred her from taking home the one they had initially supplied. Her mother, Vanessa, saw no other option in the interim but to ask her abusive ex-partner, Moira's father, for financial help to purchase a replacement. "At the time," confided Vanessa, "I was totally single. Just out of a terrible relationship. Got seriously screwed by her father. There was no way I could afford an iPad." When Vanessa and Moira moved away, Moira's new school district provided her with an iPad with Proloquo2Go that she could take anywhere, along with services to support Moira and her mom in integrating the communication device into daily life. The other iPad remained at home. "It's good," said Vanessa, "because now we have two with Proloquo2Go. If the battery runs out, then we have that one. If she's sitting next to me, trying to say something, and I just don't understand what she's saying, then I have that iPad."

How Moira came to have two iPads at home was quite different than more privileged children. Their parents tended to have owned one or more iPads prior to the child using one for AAC. The introduction of the iPad into the home often originated from an Apple device purchased years earlier by a husband in a white-collar profession. Isaac's mom, Sara, recalled, "The first iPhone came out, [and] my husband went and got an iPhone. When it was the first iPod Touch before that, he got that." Cory's mom, Perri, also highlighted the gendered nature of domestic communication technology adoption. "We are a big iPad family because my husband started with the first one," she said. "He passes it down when he gets a new one. He uses it in his business and everything." The planned obsolescence of Apple devices and a cascade of hand-me-downs resulted in an abundance of iPads in more privileged homes. Remarked Perri, "There is one at home that [my son] uses that was my daughter's, from my husband to my daughter. Now, my son has the original iPad."

By the numbers alone, most children had access to multiple iPads at home. It is important, however, to keep in mind that how these devices initially entered the household and were maintained by parents differed across class. Only certain parents read up on the newest computers on the market, let alone have the disposable income to supply their children with a steady stream of new technologies. Children of parents with white-collar jobs had the advantage of prior domestic access to an iPad long before the child might have needed another one to communicate.

Fun and Communication

More privileged parents also differed from less privileged ones in their understanding of the purposes of both devices. More privileged parents, particularly mothers with a college education or advanced degree, tended to use the emic language of "fun iPad" and "communication iPad" more so than less privileged parents. They associated fun iPads with the act of *playing* or *watching media*, and communication iPads with the act of *talking*. Because their children seemed to prefer playing to talking on an iPad, these parents sought out separate devices as a way to maximize their child's potential speech output without denying them the pleasures of playing on an iPad.

For instance, Sam's mom, Donna, said of the iPad, "You can use it for a million other things, which are a million other *good* things, but Sam can't sit there with his Proloquo2Go and play games." When I asked Eric's mom, Anne, if her son's school-issued iPad had any other apps on it besides Proloquo2Go, she replied, "That's *it*. We did that on purpose so that he couldn't play around with it." Raul's mom, Nina, described her logic for purchasing her son his own iPad expressly for AAC. "You can't just stick Proloquo2Go on any old iPad," she said. "When there's ten million apps on there, it's a 'fun iPad,' and then all of a sudden you expect the child to use it to communicate?" When given the choice, Nina contended, children "want to play the games or do whatever the apps are on there." More privileged parents tended to frame the balance between fun and communication as a justification for needing two iPads.

Issac's mom, Sara, a lawyer turned stay-at-home mother of four, reported that the family owned an incredible amount of iPads—eleven in total (ten functional and one with a shattered screen). "I can't even count the amount of Apple products we have in our house. They far exceed the family members, and we're six," said Sara. "It's sick. It's a lot of money in Apple products." At home, Isaac had a "communication iPad," "a backup for the communication iPad," and his own personal "fun one." When I asked Sara how the distinction between the fun iPad and communication iPad came to be, she declared, "You have to have one device that's solely used for your voice. Anything else that's done on there is going to take away from your ability to communicate."

More privileged parents wanted to keep the communication iPad distinct from the many media forms subsumed within the device because they might distract the child from using Proloquo2Go. Sam had started using Proloquo2Go on the iPad, but actually transitioned a few months later to the Vantage Lite, a dedicated AAC device. Donna preferred the Vantage Lite to the iPad for AAC because "the only thing my son can do with his Vantage Lite is talk. With his iPad, he can watch movies, he can play games, he can read books—he can do a thousand other things and if he doesn't push on to his Proloquo, then tough toenails." Donna now referred to Sam's iPad as his "little TV." Before giving Luke an iPad with Proloquo2Go, his dad, Rob, made sure to "lock down all the features that [Luke] would go and play with. Because YouTube was an icon on the original devices when they came in. There's no way we could have him turning that thing into a television. He wouldn't use it for anything but." To Rob, YouTube would transform the iPad into another medium altogether, so he took preventative measures to limit Luke's media consumption.

Nina, however, saw children's use of YouTube as distinctly different from their experiences with either television or Proloquo2Go. She said, "I think the distinction for me is that with the TV, you are a passive participant, right? And with the iPad—as a communication device—you are an active participant, and it provides you a meaningful way to communicate." Nina situated YouTube along a spectrum of active to passive. "I think that there is a middle of that between an iPad being a communication device and the TV," she explained. "There is YouTube, and then you still feel a bit more active because you are picking what you are interested in." While still making clear distinctions between communication and fun, Nina considered the latter category to be multifaceted, perhaps reflecting how the children's media industry positions YouTube and other Internet video-on-demand services such as Netflix, Hulu, and Amazon as "hands on" to parents.

Speech-language pathologists and school district AAC specialists also frequently framed iPads as being either for communication or fun. Noted one school specialist regarding the iPad with an AAC app, "Think of this as their mouth. What's going to come out? You don't talk in games. You talk with communication."[8] Speaking the same language as professionals was one way in which more privileged parents were able to mobilize institutionalized cultural capital in ways that less privileged parents were not. For example, in speaking about the parents and children in her caseload, Rachel told Stephanie's parents, "For most of our families, they have a communication iPad and then a fun iPad. Then it's distinguished." Bourdieu's notion of distinction plays out quite literally here, with Rachel legitimating the fun/communication classification as representative of most families. She also makes visible how the sorting of the two iPads serves as a boundary object, used and interpreted in different ways among parents and professionals.[9]

This classification ordered human interaction when Madeline's mother, Teresa, told Caren that she had been playing *Wiggles* videos for Madeline on the iPad with Proloquo2Go (which they had been temporarily lent by the regional center). Caren recoiled in slight horror and quickly responded to Teresa, "My suggestion is to keep it just for communication if you can. Some of my kids, they can handle both, but most of my kids, they've got to keep it separate. 'Cause then [Madeline's] gonna look at this as a toy or as a reward." Speech and language professionals, wary of children perceiving iPads as playthings, promoted keeping communication and fun separate.

Some more privileged parents explicitly mentioned picking this language up from professionals. Donna, for one, spent five years after her son was born working among assistive technology specialists at a regional center and had also received an advanced degree in psychology. Donna stressed, "When you have an iPad, it can do many other things. It is *not* a true AAC device. I cannot say that enough. That was drilled into my head by Caren and Susan [Caren's supervisor], and it's absolutely true."

Besides being a parent to a nonspeaking autistic child, Nina was a special education professor at a local college. She often advised other families as both a parent and expert. Nina said, "When I talk to parents about Proloquo2Go, I always say they need a dedicated communication device as an iPad. There shouldn't be any other apps on there unless they're for communication." Among more privileged parents, the communication iPad was a means of self-expression, whereas the fun iPad was a distraction. Professionals used this same language to refer to the two iPads, creating a wedge between them and less privileged parents who did not employ the same terminology.

Negotiating Fun and Communication On the Go

A few of the less privileged parents tried to make distinctions between fun and communication iPads, based on school and therapist recommendations, but found this delineation misaligned with the everyday pressures of parenting. For instance, River's parents were divorced, and he and his sister moved back and forth between their parents' homes during the week. Mark, River's dad, noted that multiple iPads were scattered across the two locations. "The first-generation iPad now lives out at Mom's," explained Mark. "That one's probably not been synced in awhile." There was also a "regular iPad" that stayed at home with Mark, and was primarily used by River's sister "for the games or Netflix." A constant throughout River's changes in location—not only across two homes, but school too—was his iPad Mini, which had "games and stuff on it" besides Proloquo2Go. Said Mark, "We're trying to get it through, 'This is a good tool for you to use.' It's hard to compete with games and stuff like that." While River's iPad Mini was "pretty much his speaking device," it served multiple purposes out of necessity.

Talen's mom, Kameelah, described her and her husband's similar dilemma in deciding "the balance" between the regular iPad that Talen had used recreationally for years and his new iPad Mini with Proloquo2Go. Kameelah remarked, "We put Proloquo on [the black-case-covered iPad Mini] and then his movies are on the other [blue-and-white-case-covered] iPad. He's got the two. His black one, he would never use for watching movies." She wondered aloud, "Do we keep his movies on the big [iPad]?" If they chose to install Proloquo on that iPad, "he's not going to want to go to that. He's going to want to go to his movies. Like, even if he did want an apple, he doesn't want it anymore because now he just wants to watch [the animated Disney film] *Ratatouille*."

Unlike more privileged parents such as Sara, who could afford multiple fun iPads for each of their children, Kameelah saw the value in keeping the option for movie watching on both devices. With little alone time available for her and husband, having films on each could keep her two young children occupied. "[Talen] doesn't realize it," Kameelah said, "but he could use the [iPad Mini] for movies. He just doesn't do Netflix yet. The icon is on that one." When I asked if the iPad Mini was used for purposes besides communication, Kameelah let out a deep sigh. "It should not be," she remarked. "It should only be used—and my husband will kill me—it should remain in one spot for him. And 85 percent of the time it does."

Having Netflix on Talen's iPad Mini, though, for use by his younger sister enabled a rare opportunity for Kameelah and her husband to spend time with one another outside the home: "Let's say my husband and I are out. This is when she'll definitely use it. When we're out at dinner and we're like, 'OK, we can stretch out the dinner and get almost like a date.' Where we can ignore them. She takes his black one, and then he takes the other movie one. And then they're using both." Kameelah placed blame on herself, not her husband, for compromising the integrity of the communication iPad classification in order to carve out more time for their relationship. She vowed soon to find a cheap alternative to the iPad Mini so that her daughter could watch Netflix on her own device instead of using Talen's. "What can I get for $150 that will let you watch Netflix or some little movies, but will leave Talen's iPad alone?" she pondered.

Explaining the difference between a fun iPad and communication iPad to a three-year-old wasn't practical either. Kameelah remarked that her daughter was "at the age now where it's hard for her to understand why Talen can have two iPads, and she can't have anything." While Kameelah did try to keep Talen's two iPads distinct, this was also harder to do in the community than at home. "In the house, it's not a big deal because obviously we have the computer. But when we're out ... ," she said, her sentence trailing off.

For on-the-move situations, Kameelah and Mark found that it was not worth keeping fun and communication on separate iPads. They agreed in theory with the idea of having Proloquo2Go and all other apps on their own devices, and did so in most cases, but parenting realities make this ideal difficult to maintain. When it comes to owning multiple devices and parenting multiple children, the decisions that parents make aren't always rational.[10] Even if they believe that fun and communication "should" belong on separate iPads, those boundaries do not always hold. These perceived breaches in respectable parenting can be a source of guilt, stress, and affective labor, especially for less privileged parents like Mark and Kameelah.

Entertainment and Education

In contrast, the iPads with Proloquo2Go used by less privileged children tended to be bought, owned, and controlled by school. Their parents often viewed the two iPads less in terms of *fun* and *communication*, and more in terms of *entertainment* and *education*.

When I asked David if he thought Beatriz used the family-owned iPad Mini differently than the school-owned iPad with Proloquo2Go, he replied of the former, "She knows that this is for play and entertainment," and of the latter, "She knows that the other one is for educational purposes." In comparison, when I asked Nina why she called one the fun iPad, she replied, "Just because it is. That's what it's for. We do have some educational apps on there, but it is for fun." Donna said, "My son's iPad now is strictly a piece of entertainment for him. I do use it for spelling tests, but that's about it." Although Nina and Donna made some concessions, they believed that any app that wasn't "for communication"—including those with educational intent—really belonged on the fun iPad, whereas David grouped school-related apps—including Proloquo2Go—together.

More privileged parents believed that any other work done in the classroom belonged on a school iPad separate from the communication one with Proloquo2Go. Sara mentioned that Isaac had a fourth iPad "for school that's just with a keyboard for typing, because it's hard for him to sit and handwrite. His motor planning is really poor, so it's just easier for him to type." The school had offered to provide Isaac with a laptop, but "we were like, 'It's a lot cheaper to just buy another iPad and have a keyboard,'" said Sara, to which the school agreed. When I asked her why Isaac needed a separate iPad from the one with Proloquo2Go, she responded, "Imagine you don't have a voice. How are you going to answer a question in class if you're expected to work on the device you're expected to talk with?"

These differing conceptions were a source of conflict between less privileged parents and professionals. Stephanie's father, Nelson, remarked that his daughter associated the school-owned iPad "with work." In fact, "The only time she prefers [the school-owned iPad] is when the batteries run out [on the family-owned iPad]." In a training session with Stephanie's parents, Rachel was openly frustrated that they had been inconsistent in encouraging Stephanie to use Proloquo2Go at home on the school-owned iPad. Nelson explained, "We figured they were using it more at school." "Don't think of this as work," replied Rachel. "This is her talking to you. It shouldn't feel like work to her. Maybe the message wasn't clear to you all that communication was the main purpose," she said in a slightly condescending tone. Yet Stephanie's mother, Marisa, was confused about communication being the primary purpose of the school-issued iPad. "But at school," Marisa interjected, "they use it for video, games." Nelson, Marisa, and Stephanie's conception of the school-issued iPad as intended for *education* conflicted with Rachel's view of the device as intended for *communication.*

When the school-issued iPad with Proloquo2Go was used at home, less privileged parents, especially those without a college degree, appreciated how their child could maximize the iPad's educational potential by using other apps on the device. Kevin's mom, Rebecca, said that his school-issued iPad with Proloquo2Go also had "his own little reading books. I think it has a math game in it, and it's got Dictionary.com, which they use" at school. She found the storybook apps to be "*really, really* helpful for him because when he sees the word *and* hears it at the same time, I think that's helped his understanding vocabulary and deciphering the sounds and seeing the words." To Rebecca, the school-issued iPad with Proloquo2Go as well as other assorted educational apps loaded on it helped her child make clear gains in literacy.

Stephanie's school-issued iPad came only with Proloquo2Go installed, but her parents went out of their way to fill it with free educational apps. "They actually gave me access to it with a password," Nelson explained.

"We've downloaded all the apps. The only one the school did was this one," he pointed out, referring to Proloquo2Go. "The girl that was from the school district said, 'Use the iPad as much as you want.'" Marisa told me a story of how one day, when she was updating the software on the family-owned iPad, Stephanie started "playing with books [on the school-owned iPad]. She start to try to read. In her language, believe me. Some words I understand. Some words I don't. But she try to read a book on the iPad," she said proudly, almost defiantly. Told by professionals that her autistic child could not read, Marisa believed that the apps her and her husband had installed on the school-issued iPad enabled Stephanie to demonstrate abilities that her teachers doubted.

Both more and less privileged parents generally viewed the school-issued iPad with Proloquo2Go as benefiting their child, and tried their best to encourage children to use the device in positive ways at home. Less privileged parents, however, valued the technology for different reasons from more privileged parents. The former group of parents tried to maximize the use of the school-owned iPad for educational purposes at home, particularly simple book and game apps from which parents could see an obvious benefit. These parents, who generally had less formal education and technical literacy than more privileged parents, saw Proloquo2Go, a complex app requiring highly specialized training, as belonging to the domain of the school.

Locking in Communication and Locking out Fun

As illustrated above, parents did not agree uniformly on the intended use of the school-issued iPad with Proloquo2Go, operating under the impression that it was either *for communication* (among more privileged parents) or *for education* (among less privileged parents). Parents who owned the child's iPad with Proloquo2Go (who also tended to be more privileged) exerted far more agency in determining how or if that device could be used in particular situations for purposes besides communication. The next section expands on this discussion, delving into how power was distributed across families and schools in the management of children's iPad use for fun, communication, and everything else.

Parents and district officials directly controlled children's access to the iPad through locking the machine in two main ways. First, the fall 2012 release of iOS 6, the sixth update to Apple's mobile operating system, included a new feature called Guided Access.[11] The setting works by temporarily disabling the use of the home button, the primary way of closing one app and switching between others. Enabling the Guided Access feature made it so that a child could be locked in to an app such as Proloquo2Go.

Second, students and parents could be locked out from certain apps on the school-owned iPad as well as the entire device at home and school. Locking out could occur when the school had turned on the Apple iOS setting requiring a passcode (a custom numeric or alphanumeric code) in order to turn on or wake the iPad, yet did not share that passcode with children and families. The power to use basic and additional features in communication technologies designated for people with disabilities is rarely held by users who might find the features beneficial, but rather by technology manufacturers and institutional providers such as schools.[12]

Locked In

The beginning of my fieldwork coincided with the introduction of Guided Access, which Apple explicitly defined as a new "accessibility" feature.[13] In his keynote address at Apple's 2012 Worldwide Developers Conference, executive Scott Forstall mentioned that Apple developed the setting based on the feedback from parents and clinicians that some autistic children have difficulty focusing on a single app.[14] There is a valid argument to be made that Guided Access should be categorized under accessibility; among users with motor control challenges, Guided Access can prevent an accidental gesture from taking an individual out of the app they use to communicate.

Yet Guided Access is also explicitly used to make other apps inaccessible. It can only be turned off when the home button is triple clicked and a preselected on-screen passcode is entered. That is why Guided Access has been described in the popular press as being for "kids and kiosks"—specifically, for locking children into apps used for school testing and enabling unstaffed iPad terminals (for instance, tablet menus for ordering restaurant food).[15] In its support materials, Apple notes that "Guided Access helps you to stay focused on a task while using your iPhone, iPad or iPod Touch."[16] Although this description emphasizes behavioral self-regulation, Guided Access is not designed to solely benefit the user; it also affords technology regulation with little direct human oversight and labor (as in the example above for restaurant wait staff).

By locking children into Proloquo2Go, Guided Access could temporarily turn an iPad with any other number of apps into a pseudo-dedicated AAC device that a child could be left with unattended. Guided Access transforms the iPad into a "walled garden"—a term used to describe children's technology made "safe" through specific limitations like Internet restrictions.[17]

Children can be quite skillful, though, at developing creative work-arounds to scale the metaphoric wall.

One school district AAC coordinator remarked that an iPad with Guided Access enabled is a dedicated device "until [students] see you enter the code." When I asked Nelson how Guided Access was working out for the family, he replied, "I thought it was an awesome idea, except [Stephanie] figured it out!" Stephanie had overheard her parents and therapists talking about the passcode, seen it demonstrated, remembered it, and then used it to unlock Guided Access. "We had to tell the people to change the passwords because in school her therapists were like, 'Don't say anything in front of her,'" said Nelson.

Caren, describing Guided Access to Daisy, explained that once enabled, the iPad "truly becomes dedicated." There was a particular moral judgment to some professionals' views of Guided Access. Rachel said it was helpful because there were some "kids who constantly get on YouTube" and "try to do everything on here but communicate to us." Rachel theorized that "[Apple] did Guided Access because people were saying, 'This is driving us crazy, watching YouTube all day.'" One school district assistive technology specialist reported that in response to the introduction of Guided Access, she'd "seen some teachers go, 'Oh good, because we don't want them in anything but communication on this device.'" While YouTube turns the iPad with Proloquo2Go into a television (per Rob), Guided Access converts it into a dedicated AAC device.

That same school district assistive technology specialist had serious concerns about teacher abuses of Guided Access too. She had seen it be helpful "in some cases where a student is resistant to doing anything on the iPad except for going to YouTube or playing a game," and "for those kids who go into the settings and mess things up, and they don't know what they did." She nonetheless had reservations about this control backfiring, explaining, "I hesitate to just lock a kid in an app and not let them get anywhere else in that iPad." She felt that such a thing would be frustrating, leading some students to "become aggressive if you don't let them out of the communication app and into the other thing." Guided Access yields more power to the person doing the guiding than the one being guided.

There were clear differences among parents in their awareness and knowledge of the Guided Access setting. More privileged parents were likely to have researched how to use Guided Access. When I asked Danny's parents, Alice and Peter, how they had first heard about the feature, Peter said, "When the new iOS was coming out, I did my research on available features of the iOS before I updated." Less privileged parents generally did not learn how to use it on their own. At one home visit, Rachel tried to enable Guided Access on Stephanie's school-owned iPad, but it required a passcode, and she wondered if the school had already set one. Nelson speculated that he might have gotten an e-mail about Guided Access from the school's assistive technology specialist, saying to his wife, "Remember we got that e-mail and thought, 'I wonder what that does?'"

More privileged parents who owned an iPad with Proloquo2Go sometimes used Guided Access as a strategic way to promote their child's communication, specifically communicating their desire to use another app. Alice explained, "We'll just triple click, and that actually starts Guided Access. That means you can't get out of this now. We can set it up so now, if you want to play an app, you have to request it." Danny could then request his spare iPad to play with by pressing an "iPad" button on his Proloquo2Go. One of the assistive technology specialists I interviewed also suggested this strategy. She said, "I've used that as communication then. Let's program in your AAC app for you to request to get out of this and access to the other thing you want. Using it that way, you've got to find what motivates them." In this manner, communication was rewarded with fun through requests to play with other apps within the locked-in Proloquo2Go.

Locked Out

Families that did not own the iPad with Proloquo2Go were regulated by the school in how their children used the device at home and for purposes other than communication. Thomas's mom, Daisy, did not agree with the school's presumption that her son would always favor playing around over using Proloquo2Go. She had initially thought that Thomas could not handle using the iPad for communication with other apps installed. "In the beginning," she recalled, "I don't think it would have been a good idea because then he would just skip the communication ... and go to the games."

Eventually, however, her son's desire to communicate with Proloquo2Go grew and could override his interest in playing with other apps. "He knows that if he didn't use it," said Daisy, "we wouldn't give him anything that he requests and asks for, or listen to his phrases, comments, what he feels if he doesn't use this." While Daisy may have thought that her son could manage toggling between Proloquo2Go and other apps, Thomas's school district had gone ahead and preemptively locked him out of using the school-owned iPad in any capacity other than for communication.

The fact that the school "took out a lot of stuff," including the web browser, angered Daisy for three reasons. First, she thought that it was not reflective of how iPads are actually used by the general population. According to Daisy, "In the *real world*, it would make sense if it was all on just one device. That, OK, 'If I wanna communicate with you, ta-da.' And then, 'Oh, I feel like playing a game,' then it's gonna be the same." The school had instead forced the fun/communication classification on Daisy: "'Oh, lemme do this [on one iPad] for the game,' and then 'Oh, lemme do this [on another iPad] for the communication.'" "Real-world" application of skills was important to Daisy. She employed an after-school aide to take Thomas on trips to the grocery store and post office to foster community participation. To Daisy, having a communication iPad separate from another iPad was not practical beyond the walls of Thomas's classroom.

Second, Daisy felt the school was not allowing her son to take full advantage of the technology's capabilities. She explained, "If you're nonverbal, you would use this to communicate, and then, 'Oh, now it's school. Let me use this to write my notes.' Or 'use this for my school research.' Or, 'Oh, let me check my e-mail.'" Instead of using the iPad for one purpose, "You would want it to be all in one," said Daisy. She was frustrated that "the school is limiting it to, 'Nope. This is a communication device. That's it.'" In Daisy's eyes, its full potential was being wasted. Some professionals also shared this view. One noted, "As an assistive technology specialist, I feel that we should get as much out of [the iPad] as we can, and communication can be a great piece of it, but it doesn't have to be the only piece." Those aspects included "interaction or communication or learning their ABCs or whatever it is. If their eyes light up when they're doing it, then use it."

Lastly, Daisy viewed the district's control over the iPad as symbolic of an imbalance of power. "I don't know how the other school districts are," Daisy reasoned. "I shouldn't complain too much. Just the fact that they gave me an iPad, and the fact that I'm able to put stuff [in Proloquo2Go] for home stuff, and that it's just his and that we get to take it home." Daisy was grateful for what the school district had given her son, especially considering that their assessment and delivery process for an AAC device had been drawn out over years.

Daisy was right: things could be worse. While Guided Access could lock a child into an app, the school and its technological infrastructure could lock them out completely through the passcode setting on the iPad's lock screen. Passcodes can be set, used, and changed on any Apple device. The company expressly describes the passcode in terms of personal security, noting on its support page, "Set a passcode on your iOS device to help protect your data."[18] Still, storing personal data on individually versus institutionally owned devices (e.g., iPhones given to employees or iPads given to students) highlights an inherent tension between privacy and security.

While an in-depth discussion of cybersecurity is beyond the scope of this book, one need look no further for evidence of this friction than Apple's February 2016 objection to a California court order for the company to create software enabling it to bypass any user's passcode. Federal Bureau of Investigation (FBI) officials initiated the request in order to access the contents of the iPhone used by the deceased gunman in a December 2015 terrorist attack in San Bernardino, California.[19] The iPhone in question was issued by the gunman's employer, the San Bernardino County Department of Public Health. The county complied with FBI requests to reset the password on the iCloud account associated with the phone's data storage, but this reset inadvertently rendered the phone unusable unless the device's passcode, ostensibly known only to the gunman, was entered.[20]

Tensions emerge when institutions, be they employers or school districts, purchase and own "personal" mobile technologies. Users tactically respond to the constraints of device affordances, markets, and policies. On the day of my second interview with Beatriz's parents, for instance, her school-issued iPad came home newly passcode protected. There had been no message sent from the school that this change would be made. It was unclear if the enabling of the passcode resulted from some kind of automatic or inadvertent software update over the school network, or a manual change in device settings. In any case, the passcode had not been shared either with Beatriz or her parents. Her father, David, remarked, "She never had a password on that iPad. They put it there yesterday, and today again. So I cannot see."

Beatriz took the iPad from David's hands and pressed a combination of numbers on the lock screen. The system interpreted these interactions as too many purposeful but unsuccessful attempts to enter the passcode. Without warning, Beatriz was locked out of the iPad for an hour. I took a photo of the iPad screen with my iPhone (figure 4.1). At the time, it read, "iPad is disabled; try again in 54 minutes." Ironically, not only was the mobile communication device "disabled" in this situation (presumably by Beatriz's own attempts at unlocking it) but the district also disabled Beatriz in her use of the iPad. This sudden change in access was likely in violation of Beatriz's IEP, which guaranteed her the right to communicate with the device, yet David was hesitant to challenge the school's authority for fear of

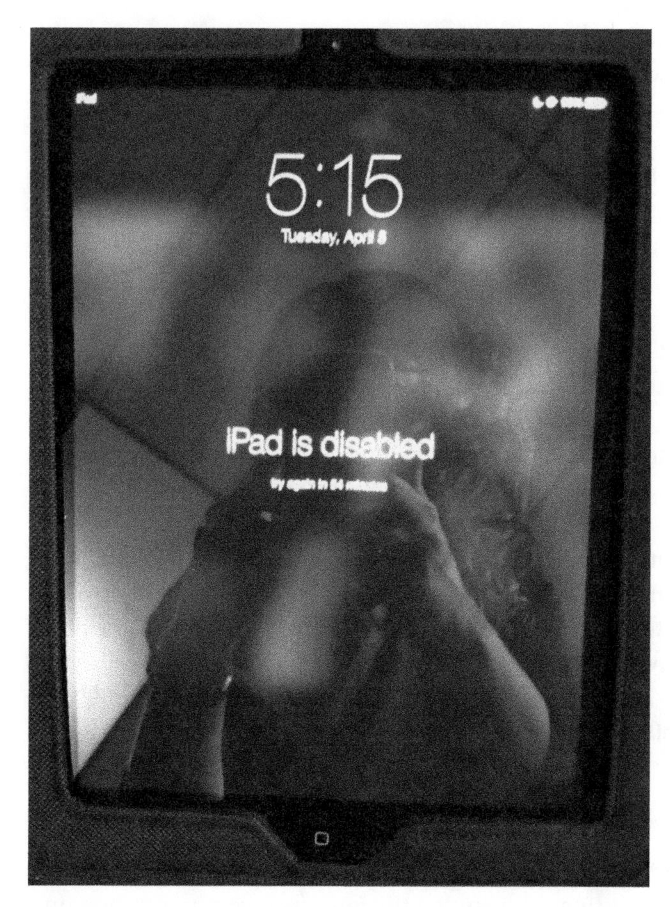

Figure 4.1
iPhone photo taken by the author of the screen of Beatriz's school-issued iPad, which reads, "iPad is disabled; try again in 54 minutes."

retribution. He confided, "I'm always afraid that if we go against the school district, they may take measures against her, like disliking her."

Beatriz was free to transport the mobile device back and forth between home and school, but both the device and her communicative agency were immobilized by the school's device settings and technology management infrastructure once she walked through the doors of her home. Along the way, both humans and machines made decisions that rendered Beatriz's iPad simultaneously, and irreconcilably, mobile and immobile. The clock on Beatriz's iPad screen continued to countdown as I left her home that day. She could conceivably "try again" after an hour had expired, but what would have been the point? The iPad would not be reenabled until Beatriz brought it into school the next day—the immobile mooring for her mobile device.

One district AAC specialist explained that "if a family by mistake plugs the device into their computer, it will lock the iPad. You have to bring it to the 'mothership' [the nickname for a hub in a technology management building in the school district] to get it unlocked or to do any of the software updates." To prevent this, AAC specialists in the district would "put tape all over" for parents that read, "'Do not plug in please.' Whatever you do, don't install software updates." The purpose of the mothership, as nefarious as that name sounds, was not to arbitrarily control the child's iPad but instead to protect the school's financial investment. When I asked the AAC specialist about the logic behind the self-disabling iPad, she responded that "[parents] could potentially delete the app and then we would lose the license for the app."

Far from a simple emancipatory tool, then, for giving voice to individual users, institutionally owned Apple devices are bound to external security and financial interests. While networked iPad and iPhone "locks" are partly created by hardware and software settings, they are also forged through sociocultural, political, and economic factors—in this case, through the intersecting political economies of educational, mobile, and assistive technology.

Unlocked

Prior to the iPad, it was not as if users of dedicated AAC devices or their parents needed to distinguish between the fun versus communication one. A dedicated AAC device like a Dynavox speech-generating system has one primary purpose—to aid oral speech—and so it is not a technology that families would otherwise have purchased or own two of. At the same time, even a "dedicated" AAC device can have a plethora of secondary functions. Donna raved about the features of the Vantage Lite AAC system, saying, "It's a bunch of other things too. It's a telephone; you could call people on it. You can use it for a remote control for the TV. … It's a pretty advanced device." "Dedicated" AAC device is a bit of a misnomer, as these technologies are used to store, circulate, and manage various types of media.

A key difference between a multipurpose iPad and multipurpose Dynavox is that with a Dynavox, those secondary functions are hidden unless otherwise unlocked—which comes at a price. "When insurance pays for [a Dynavox]," explained one school district AAC specialist, "it comes as a locked device. It will come with only access to the communication component." For an extra fee, companies will unlock these features. "It's about $200," the specialist said. "If the family pays out of pocket, the company will send them a code or a USB drive, and that will unlock the other features of the system." Insurance companies lock users who cannot afford to pay into a definition of what it means to use a mobile device to communicate and refuse to cover iPads as AAC devices. "Communication is e-mail, communication is text messaging, it's FaceTime, it's all these other features that are available, but we have a very narrow definition of it," the AAC specialist contended. "I think it's evolving, but it's very slowly evolving."

Learning to Speak the "Language"

The field of AAC is slowly recognizing that communication is multifaceted, but insurance policies to date stop short of reflecting how people use new media to meet their daily communication needs. Changes in 2015 to Medicare coverage for speech-generating devices expanded the definition of "speech" to include technologies capable of communication by text, e-mail, and telephone. The stipulation is that those devices must also meet the definition of "durable medical equipment," which automatically excludes personal computers, tablets, and mobile devices.[21] Medicare policy also equates these modes of communication as akin to "functional speaking," while excluding software used to create documents, play games, or chat via video.

Communication, write AAC researchers Janice Light and David McNaughton, also extends to "the development of social relationships, the exchange of information, and participation in social etiquette routines" through social media and pervasive communication tools, such as e-mail, texting, and blogging.[22] In the coming years, it will be increasingly difficult for professionals and policy makers to isolate AAC apps on off-the-shelf mobile technologies from other forms of electronic communication that allow nonspeaking individuals to interact with, participate in, and influence society and their social surroundings.[23]

The extent to which communication technologies can benefit non- and minimally speaking individuals depends on having agency to define their own communication needs. This negotiation over the terms and terminology of communication has a long history. For example, in her analysis of telephone-based home communication alarm systems for individuals with disabilities, Patricia Thornton found that the manufacturers and providers of said technologies imposed rules about appropriate communication.[24]

While users were expected to only utilize the service for emergency response and not to "abuse" the system, individuals desired to use it for daily living in order to stop emergencies from ever occurring (e.g., preventing falls by requesting assistance getting in and out of bed).

Paternalistic approaches to providing technology for children with communication disabilities limit the potential for young people and their families to determine how the iPad might best suit their needs. Whether or not children's iPads with Proloquo2Go should be multipurpose—for fun *and* communication, or entertainment *and* education—is a complicated question. Yet the answer should rightly center on the preferences and abilities of the child using the iPad. Some young people may benefit from exclusively having access only to Proloquo2Go, and others may be capable of toggling between apps.

The debate about "proper" individual iPad use is important, but also has the potential to distract us from discussing broader structural issues—namely, on what grounds decisions are based—and the unequal distribution of decision-making power among more and less privileged parents. It is disconcerting that the use of the iPad as a communicative tool is more likely to be controlled by school districts among less privileged children. Districts have other priorities besides children's individual freedoms, such as protecting large investments in technology and technological infrastructure.

The degree to which children could utilize the school-owned iPad for learning at home, outside Proloquo2Go, was highly variable too. Speech-language professionals chastised Stephanie's parents for mining the learning opportunities of the iPad. Daisy and David felt let down and confused by their experiences with the district, which placed limits carte blanche on how or if their children could use the school-owned technology at home. A number of digital media and learning researchers have focused their work on "connected learning," or the potential for networked communication and new media to bridge children's learning opportunities in and outside school.[25] Daisy, David, Nelson, and Marisa's stories illustrate how these connections are delicate. In their own homes, children were always tethered to the school's management of the software and operating system, for better and worse.

In using the terms "fun iPad" and "communication iPad," more privileged parents demonstrated a particular "habitus," or way of making meaning in daily life.[26] The idea that an iPad should only be used for AAC is learned. This hinges on parents' language fluency, educational background, technical literacy, and free time. There is nothing inherently socially or culturally stratifying about two groups of parents viewing two iPads differently. Habitus occurs, though, in a *field*, or social and cultural arenas of human action and interaction.[27] In this study, privileged parents were advantaged in their dealings with clinicians, who also tended to use the language of fun and communication, and looked more favorably on parents who viewed the iPad similarly. Parent participation in education requires the mobilization of specialized kinds of cultural and social capital.[28] Speaking the same language as experts further advantaged parents who generally encounter less structural barriers to participation in their children's education.

Conclusion

While both more and less privileged parents mostly understood their child's two iPads as having distinct purposes, they distinguished them differently. More privileged parents framed the duo of iPads in terms of fun and communication, while those less privileged understood them as being either for entertainment or education. While both the fun iPad and communication iPad are materially identical communication technologies, parents who employed the euphemisms only considered the latter to truly be a technology for communication. In other words, fun iPads and communication iPads are symbolically different communication technologies for parents who use this language. Less privileged parents did not care *less* about their child benefiting from the iPad with Proloquo2Go—they just cared *differently*.

More privileged parents, like Sara, had many more iPads, and so could dedicate one to be the communication iPad. Less privileged parents, like Kameelah, owned far fewer and at times needed the device to do double duty; they did not have the luxury of creating a neat binary. This difference between parents highlights the role of objectified cultural capital, as there are class biases inherent in professionals recommending that the child's iPad should only be used for AAC when only the most privileged parents have multiple iPads for each child.

Children's ability to use the iPad for various purposes was bound up with overlapping power structures. The political economies of assistive technologies and educational technologies both serve here to potentially increase the gap between what children need from the system and what they receive. While health care policy and insurance companies treat assistive technologies and communication technologies as separate (the former as durable medical equipment and the latter not), parents' social construction of the fun iPad and communication iPad illustrate how assistive and communication technologies coconstitute one another.

Instead of seeing fun as separate from communication, parents also discussed it as fundamental to expression. In the next chapter, I explore how some parents, across class, understood their children's recreational iPad use (e.g., watching YouTube videos or playing with the video game Minecraft) as inherently communicative as well. This more open definition of communication, however, opened more doors to better resources for more privileged families.

5 Augmenting Communication with New Media and Popular Culture: What Does It Mean to Communicate with an iPad?

> Technology's saved [Danny] from a diagnosis of autism *and* retardation because we were in with a doctor who was about to diagnose him as retarded and autistic. He was being a pain in the neck, and I handed him my iPod, and he opened up the iPod, moved two screens over, opened up videos, looked at them, and opened the video he wanted, got bored, and went and played a game. You could see the doctor just sit there and go, "Not retarded."[1]
>
> —Peter, Danny's father

In industrialized nations, some form of mobile media almost always accompanies caregivers and children during the course of their daily activities, be it a phone, tablet computer, gaming device, or music player. Trips to the grocery store, visits with relatives, and appointments at the doctor's office—all expand the social context of children's engagement with media and technology. While young people may be physically colocated with their family members as they engage with mobile media, the screen is often characterized as intruding on the family unit (media as "babysitter") or controlling children's minds (media as "addictive").[2]

Peter's quote above illuminates another perspective. In my interview with him, he bluntly expressed the belief that Danny's iPod navigation in his pediatrician's exam room quickly relayed a degree of intelligence to the doctor that Danny ostensibly could not have performed through talking, and did not already measure up to using established medical and psychological evaluations. Based on his cultural knowledge about Apple technology, techniques for operating touch screen devices, and rituals of media browsing, the pediatrician reportedly used his observation of Danny's intentional and purposeful iPod use to casually appraise the boy's level of cognition.

In the previous chapter, I detailed how more privileged parents and speech-language pathologists tended to treat "fun" uses of the iPad as separate from "communication." While an iPad has the capacity to be a technology for information and communication in a broad sense, these parents and professionals preferred that nonspeaking children dedicate at least one iPad to the production of synthetic speech, reserving the use of all other apps for a separate iPad. A number of both more and less privileged parents, though, complicated this fun/communication binary. I found many parents, like Peter, who considered their nonspeaking child's recreational media use to "say something" important about them. They understood their children's recreational engagement with new media—besides the Proloquo2Go app—as inherently communicative.

This chapter examines how parents interpreted their non- or minimally speaking child's engagement with new media and popular culture in various settings and under different circumstances as an expression of the very skills in which the dominant able-bodied culture frequently presumes them to be deficient. I begin by reviewing two areas of research that help situate these findings. First, drawing on disability studies, I explain challenges to the deficit model of disability, which characterizes nonspeaking individuals as lacking in communication skills.[3] Second, in light of work in language and literacy studies, I detail tensions surrounding a deficit model of children's recreational media use, in which popular culture is defined by its lack of value for children.[4] I next outline three areas of aptitude—socioemotional, cognitive, and verbal skills—that parents discussed emerging through their children's use of communication technology, including the iPad. Given the dominant perceptions of both disability and children's popular culture as deficient, I consider the potential of more enabling asset-oriented models. This research, I conclude, not only has implications for youth with disabilities and their families but also for the normative study of children and media writ large.

Communicating Deficiency

There are various competing logics for explaining disability as a concept. The dominant ideology in the United States is the deficit model of disability, also known as the medical or individualist model.[5] The deficit model emphasizes what people with disabilities are thought to lack, and this absence becomes their defining trait.[6] The goal of those operating under the logic of deficit is to repair the broken disabled body, make whole the incomplete disabled body, or normalize the abnormal disabled body.[7] In the deficit model, disability diminishes an individual's life; a life without disability is set up as the standard against which individuals with disabilities are compared.

For example, the language of deficit dominates autism discourse.[8] This rhetoric characterizes nonspeaking autistic individuals as lacking in socio-emotional, cognitive, and verbal skills. Clinicians in the United States tend to label autistics who can speak as "high functioning" and those unable to produce embodied oral speech as "low functioning." Medical and scientific autism experts maintain that autistics are unable to employ a "theory of mind," or comprehend the feelings and intentions of others.[9] In the 1980s, psychologist Simon Baron-Cohen advanced the idea that autistic individuals are "mindblind," causing them to seek social isolation.[10] An inability to talk or respond quickly to questions is often mistaken for intellectual disability, as evidenced by Danny's near misdiagnosis.[11]

Such a conception of autism and speech impairment is value laden; it dismisses the needs of those viewed as high functioning and the strengths of those viewed as low functioning. Mindblindness discounts the lived experiences of autistic people and privileges scientists with the rhetorical power to define autism.[12] John Duffy and Rebecca Dorner write, "The theory is, in effect, 'mindblind' with regard to autistic perspectives."[13] An autistic person might view behavior that Western culture tends to characterize as impolite or antisocial, such as not speaking or making eye contact, as part and parcel of being true to oneself.[14] The inability for others to relate to and communicate with nonspeaking autistics is rarely itself seen as a deficiency in the dominant culture.

Mass media increasingly allow nonspeaking individuals to provide first-person accounts of how they experience the world.[15] There is a growing genre of books in English written or cowritten by nonspeaking autistic youth and young adults: *The Reason I Jump* by Naoki Higashida, *Ido in Autismland* by Ido Kedar, and *Carly's Voice* by Arthur Fleischmann and his daughter, Carly.[16] Their work illustrates how the supposed-incapacities of nonspeaking individuals are not an objective truth but rather shaped by society and culture. A significant shortcoming of these memoirs is that they have largely been limited to representing white, middle-class, US childhoods, and should not be universalized. A more diverse range of firsthand accounts by nonspeaking youth and their families would open up further possibilities and perspectives.

Deficit Model of Children and Popular Culture

Adults often characterize children's media use in terms of a different sort of deficit model.[17] Educators tend to see popular culture as distracting children from worthwhile learning and displacing time spent with activities more aligned with a middle-class milieu, such as book reading, practicing music, and going to museums.[18] Mass media (e.g., comics, television, and video games) are thought to lack wholesome nourishment; consumption should be curbed, just as too much junk food causes young children to be vitamin deficient.[19]

This deficit view is predicated on the assumption that children lack the capacity to resist commercial messages, compared to the adults they have yet to become, also known as a model of oversocialization.[20] Appraisals of children's popular media are grounded in adult critiques of mass-produced culture and condescension toward "the masses" for being susceptible to the illusions of the culture industry.[21] Cultural studies scholars argue that this underestimates the agency of audiences.[22] Anthropologists have also long noted how the material of popular culture becomes a resource we draw on to make sense of our daily lives.[23] Media texts provide a shared set of meanings and experiences for children and their social partners to call on in constructing as well as maintaining a sense of reality.[24]

Media, outside purely "educational" content, can be particularly meaningful for youth with disabilities and their families, since they may be socially excluded in other areas of their lives.[25] Popular culture provides contexts for families to stay connected at home and at a distance.[26] Families of children with disabilities can enjoy and benefit from shared consumption and creative appropriation of media content.[27] In her ethnographic study of African American children with significant disabilities and chronic illnesses, for instance, Cheryl Mattingly found that the narratives and characters in Disney films offered rich material for children and their families to socially construct personal identities as a form of resistance against otherwise stigmatizing labels.[28]

Instead of viewing popular culture as a deficit, then, new literacies researchers increasingly understand media as a potential asset to children's development of communication skills.[29] Muriel Robinson and Bernardo Turnbull write that "what might in deficit models be rated as either irrelevant or damaging (the Disney film, the computer games)," in an asset model is "drawn on in an equally wide range of communicative practices and events."[30] These communicative practices and events consist of the array of activities through which children learn to make sense of the texts and technologies in their environment.[31] Among nonspeaking youth, the application of this asset model requires expanding the range of what activities and behaviors are considered to be communicative as well as a critical examination of power differentials in determining and diagnosing communication deficiencies.

Communicating through Media Use

The following section details how the aforementioned angles on deficit—that disability is defined by shortcomings, and popular culture causes child deficiencies—were challenged in how parents spoke of the communicative value of their nonspeaking child's recreational engagement with media. For nonspeaking autistic youth, characterized as significantly lacking in empathy, this meant communicating social and emotional skills through their media use. For nonspeaking youth diagnosed with an intellectual disability or suspected by doctors to have one, this meant communicating their cognitive abilities through their interactions with new media. For nonspeaking youth, generally characterized as "nonverbal," this meant communicating their interest in print culture.

Social and Emotional

A number of parents described their nonspeaking autistic children as social and full of complex emotions. Media play a key role in parents' stories as tools for enabling their children's expressions of empathy.

Moira When Moira (age ten)—a white autistic girl—was young, her great-grandmother, Gigi, would often take care of her while her single mother, Vanessa, went to school or worked part-time. Vanessa remarked that Moira "had a very close relationship with her great-grandmother," who had passed away three years ago. Vanessa recalled how Moira and her great-grandmother would regularly "sit on Gigi's recliner, and they would watch *Judge Mathis, Judge Judy, People's Court.*" When I asked Vanessa about what YouTube videos Moira currently liked to watch, Vanessa said that clips of those particular reality court shows were Moira's favorites. She "will also watch the commercials that they show during this time. Like 'Injury Attorney, Larry H. Parker,'" noted Vanessa. "I joke with Moira that, 'You're a lonely fifty-year-old housewife, Moira, watching your court shows, and your attorney commercials.'"

While cultural critics might deem such viewing as age inappropriate and lowbrow, Vanessa indicated that these shows provided Moira with a valuable and pleasurable experience. She believed that while Moira was unable to speak about her connection with her grandmother or potential grief over Gigi's death, she expressed their emotional link though ritualistic YouTube viewing. Said Vanessa, "It's just really amazing that she remembers that, those commercials. She was a baby and toddler. We don't watch TV. It's really endearing that she does that, because it's totally tied to her great-grandmother." Vanessa had ruled out the possibility that Moira encountered court shows on television at an older age; the family did not have cable and only used the television set to watch DVDs. Vanessa intuitively thought that by watching the television shows and commercials that Moira and Gigi shared together, Moira was re-creating comforting memories.

Stephanie Stephanie's mother, Marisa, was accustomed to hearing from professionals that her daughter lacked empathy. Sitting at the family's kitchen table, Marisa said, "Sometimes I talk with [Stephanie] or try to understand what she say. And I talk. 'Yes, mommy. No, mommy.' It's like trying to have a conversation with her." When she told one of Stephanie's therapists about these mother–daughter exchanges, Marisa revealed that she was treated condescendingly and even accused of lying: "She told me, 'Ah, Mrs. Hernandez, I love how you try to understand your daughter, and maybe I don't understand, but you made that up.'" Marisa sighed heavily during the interview and continued, "Well, I try to have a conversation with [Stephanie], and try to talk, don't ignore her. ... Stephanie give you opportunities to understand her and see in her world. It's up to you if you took that opportunity."

Marisa felt that therapists dehumanized her child and others like her. "They think they don't have feelings, they don't able to do things for you. That they are a separate person, like you and me, because they cannot express," said Marisa. Professionals lost sight of the "human part," Marisa explained. "Sometimes, I feel [it] is only, one more case, one more child. They don't really want to help the kids. They're just, one more." The situation Marisa described was unfortunate for all. The professionals who encountered Stephanie were unwilling to share in her world, Stephanie's agency was denied, and her mother was belittled.

Stephanie's world, Marisa commented, was deeply emotional despite not being able to express those feelings in spoken words. "For us," she confided, "it's a little bit bitter to understand how [Stephanie] feels. Sometimes she cries." Stephanie's engagement with media, both alone and with others, provided "a little window" onto her emotional states. Marisa explained, "We start to notice about music. Sometime when she feels sad, she put sad music [on] and she cry. When she feel happy, [it] is like the way she say, 'I feel happy.'" While Stephanie's therapists privileged oral speech, Marisa felt that Stephanie voiced her emotions through the songs she chose to play.

Marisa also experienced firsthand Stephanie's profound capacity for empathy. About an hour and a half into our two-hour-long interview, Marisa told me the following story:

> Sometimes I don't feel good. Sometimes I feel very tired or frustrated. But many times I start playing with her. Stephanie have very nice heart. The other day, I don't feel good. I'm in remission. I have cancer, so sometimes I feel tired. The other day, I don't feel good, and I sit down on the sofa. Every time Stephanie look me like that, she sit down with me, and she put a blanket on top of me. We watched TV until Nelson come home. So that part is, "I'm going to take care of you, Mom, because you take care of me."

My heart ached when Marisa first told me this story, and her account does not feel any less raw each time I read it. Marisa's cancer had not come up earlier in the interview, and only after she spoke of it, in passing, did I notice the port scars from chemotherapy beneath her collarbone, framed by the zippered edges of her gray, hooded sweatshirt.

The simple moralistic framing of media consumption as passive and therefore bad does not apply here. The television provided needed physical passivity for Marisa—an escape from draining chemotherapy treatments. Directly gazing at her mother, placing a blanket atop her, and sitting down with her to watch television—all were active expressions of compassion by Stephanie. To Marisa, Stephanie's actions communicated both her daughter's acknowledgment of Marisa's own caretaking role and Marisa's need to be cared for. Said Marisa, "You think she don't understand you, but she do. She do." In spite of naysayers, Marisa and Vanessa believed that their autistic daughters displayed vast emotional intelligence, and their intimate social experiences with media were proof positive.

Cognitive

Parents of nonspeaking children, including those with and without a diagnosis of autism, interpreted their child's mastery of media settings and controls on the iPad as communicating an aspect of their intelligence discounted by others. Gesture plays a significant role in communication and development, with physical gestures (not only sign language) a form of utterances made visible.[32] At the beginning of the chapter, I quoted Danny's dad, Peter, who described how witnessing Danny's iPod navigation in the exam room single-handedly convinced Danny's doctor at the last minute not to diagnose him as "retarded." A number of parents similarly explained how technology enabled their children to show that they were more intelligent than developmental experts gave them credit for.

Beatriz According to her father, David, the manner in which Beatriz played with computer games at home offered key insights into her memory and comprehension. "There are very challenging games that she plays and shows how smart she is," said David. He gave one example of a game in which the player is presented with a grid of four items: three of the items share some quality in common, and the player must identify the one that does not belong. Beatriz easily mastered the game, much to David's surprise. He remarked, "I just look at her, from far, and try to pay attention to what she's doing, to see if she's doing it by *chance*. But she's doing it constantly. That shows me that she really knows what she's doing."

David also detailed how, through online games, Beatriz displayed a keen ability to understand musical patterns. She frequently played games on the children's educational website Starfall.com and would hum along with the accompanying music. David said that Beatriz had the ability to recognize the rhythm, start and end the phrasing at the right time, and anticipate the conclusion of the music. "That tells me that she *knows* the song. She cannot sing it, but she knows it," he asserted. The inability to produce embodied oral speech did not preclude Beatriz from having an innate sense of what singing was like.

In addition to games, David interpreted Beatriz's YouTube use as evidence of her cognitive skills. As with the computer games, David initially thought that Beatriz's interactions with the website were not intentional. Recalled David, "At first, we thought that she was just watching items because that's what her finger picked. But then we noticed that she's constantly choosing the same items, so that tells us that she *knows* what she wants." Beatriz could not only purposefully choose videos but also work around YouTube's automated algorithmic choices:

> She knows where the History button is, and after she's been redirected several times to something that she doesn't like anymore. ... YouTube, little by little, is redirecting you to something else. When she goes to that extent, then she presses the History button, and then—I don't know how—but she finds her video. Right? I don't know how she does it. I don't know if it's the length of the letters, the phrase, the length of the title, I don't know, but she finds it.

David gained a new appreciation for his daughter's cognitive abilities through her use of online games and video sites. He interpreted Beatriz's intelligence not through hearing her words but instead observing her repeated patterns of action and interaction with new media.

Talen During part of my interview with Kameelah, her son Talen (age six)—a mixed-race autistic boy—sat nearby repeatedly rewinding parts of the Disney movie *Toy Story* on his iPad. Kameelah explained that Talen frequently manipulated the iPad interface in this manner. "I didn't realize you could rewind on an iPad ... so I watched him do it," said Kameelah. "He watches it and rewinds sometimes, and lets it go, watches it and rewinds, lets it go. He goes to different shows and has his parts. He'll fast-forward to the right scene he wants. Rewind to the right part." In his manipulation of the interface, Talen taught Kameelah something new about the technology.

I told Kameelah how I had observed a number of autistic children engaging in the same sort of activity, and that research seemed to confirm this behavioral pattern.[33] I asked her, "Do you have an idea why he likes to do that?" Kameelah suspected that it provided Talen with sensory pleasure. "A couple of things I've watched on *Toy Story*, he likes what they're doing, the motion of what they're doing. Other parts ... he likes the sound of it. ... I imagine it's a sensory something." She observed that Talen also liked to hold the iPad up and bring the speaker close to his ear. "I don't know if it's you feel you want to hold it because you want to feel it," she said, "because he also likes things really loud."

She emphasized how this navigation was purposeful and reflected Talen's cognitive abilities and proprioceptive awareness. "He's not following it for the story line. He's getting something else out of it," she said. "Somewhere in there, he's hearing someone. ... I think for us seeing that helped us know that there's more going on in there." In fact, when Talen's former doctor suggested that he might have an intellectual disability, it was Talen's ability to navigate the video interface with precision that, for Kameelah, cast doubt on the doctor's implication:

> His developmental pediatrician—*who we left*—she said to me, "At some point in time he should be tested for an intellectual disability," or whatever. I get that obviously something is skewed to the left with little man. But I don't ... because of different things that I see him doing, I don't know that I would say that he's intellectually disabled or whatever that might mean. ... I figure he just can't do things in the *traditional* way of doing them, but there's got to be something else

> there. Watching him manipulate that [iPad video interface] has made me like, "*Huh*. Something about that [manipulation] that makes sense for you."

Countless times, experts told Beatriz and Talen's parents that their children were not cognitively "there." More important, Beatriz and Talen likely heard those damaging words too. Their parents were a bit puzzled by their children's ability to learn new technologies, but also pleased. Their dexterity with new media communicated an unspoken yet ever-present level of intelligence and awareness.

Verbal

There were a number of parents who emphasized that although their child did not speak, they had a deep interest in words. While the term "nonverbal" is used clinically to describe children with significant difficulty producing embodied oral speech, parents portrayed their children as highly verbal in the sense that they enjoyed reading and playing with texts across multiple media.

Cory When I first sat down for an interview with Cory's mom, Perri, I began by asking her to tell me the "origin story" of how Cory (age four) came to use Proloquo2Go on the iPad. Far longer than other interviewees, Perri talked for twenty minutes straight without interruption. She spared no detail in describing the medical issues her son faced due to a combination of rare genetic conditions, developmental disorders, and a traumatic brain injury. "We have no idea what he will be capable of and what will control what," Perri remarked frankly.

While Cory had been making some speech approximations like "mama" and "dada daddy" up until age two, one day, Perri recalled, "everything left. I noticed like, 'Wow, why doesn't he say that anymore?'" With a cadre of therapists, the family tried out many AAC iPad apps with Cory, but the only one he took to was Proloquo2Go. Perri largely attributed Cory's success to how his speech-language pathologist had customized the app to take advantage of Cory's verbal abilities. He had been using Proloquo2Go "without any icons, only words, because he's so very strong in reading," Perri said proudly.

At the end of Perri's long introductory story, my first follow-up question was for her to tell me more "about reading and the *way* that he reads. I'm curious, does he use the iPad as a communication device as he reads?" I was curious if the iPad helped Cory to speak words aloud as he read, since reading aloud often helps early readers with comprehension.[34] I was struck, though, by Perri's interpretation of my question. She responded, "So the way I know. ... There's many ways that I know that he can read."

I thought that I had asked Perri about *the way* or manner in which Cory read; Perri thought that I had asked her about *the way she knew* Cory read, or for proof that Cory could, in fact, read. Perhaps she misheard me. Perhaps she interpreted some other nonverbal cue—my tone of voice, facial expression, or body language—as skeptical. Or perhaps Perri had regularly encountered others who doubted that her son with multiple disabilities was already reading at age four, and those encounters shaped her reaction to my question.

While I cannot explain Perri's response with any certainty, reconstructing her defense of Cory's reading abilities clearly reveals that various media play a significant role in how she knows Cory can read. She noted that he is "obsessed with letters. He always has been since he was an infant." While Cory has significant difficulty talking, "he tries to read books like *Little Blue Truck*. He'll try to say all the words. ... You can't really understand him. Slowly but surely, you can understand him more." Cory did not use the iPad to read aloud but rather attempted to speak the words on the page without the assistance of a speech-generating device.

His love of music was another indicator. "Another way is he knows his alphabet, and he says his alphabet all the time. We have lots of alphabet songs," said Perri. His dexterity with and independent use of literacy apps on the iPad was an additional sign. She gave an example of a spelling app Cory liked that would show a picture of an object and play an audio file of the spoken word. "Then it will give you the letters down here, and you have to drag them into the squares. I don't even have the color association help or the letter association help [enabled]. He can drag them all up there," explained Perri.

Cory's interest in television shows and DVDs with a literacy-centered curriculum was particularly strong. "He always likes *Blue's Clues Alphabet* and *Barney Alphabet*. Now he's obsessed with *WordWorld, Super Why*, all these shows that are very supportive of reading and the alphabet." Perri discovered that it was not the *Blue's Clues* or *Barney* series that fascinated Cory but rather only the literacy-specific episodes. "I would try different [*Barney* episodes] with him because I thought, 'Oh, *Barney*. He likes *Barney*.' No, he would not," said Perri. "The only one that would captivate him is *Barney ABC*." The PBS kids' series *WordWorld* was Cory's favorite television show. When I visited Cory's classroom with Perri as part of our interview, I saw that he had a red *WordWorld* lunch bag with his name on it. Perri had also downloaded all of the episodes for him since it no longer airs regularly on PBS.

Overall, Cory's engagement with literacy-oriented media was the primary explanation Perri provided as evidence of her son's verbal abilities. "I don't even know how I can tell [that he can read]," she stated, "but I could tell by the way he interacts with me and with any verbal thing or a word, word program, or words, with his books, with the cards, when he looks at words, and he tries to say them." To Perri, Cory says that he is a child who reads not through reading aloud but rather through his mastery of and interest in print, audio, visual, and digital media that focus on early literacy.

Kevin Like Perri, Kevin's mom, Rebecca, also described her nonspeaking son—who is thirteen and autistic—as deeply interested in the textual world. "I think he's kind of fascinated by words, the look of words," Rebecca said of Kevin. "It's curious because he's so language delayed." Similar to Perri, Rebecca drew heavily on examples of her son's media use to communicate to me his verbal abilities. Rebecca indicated that her family did not have cable television (due in large part to its cost), but did have an extensive DVD collection. When I asked her if Kevin had any favorites, she replied, "He used to love *Harry Potter*. He watched that over and over." As for why that movie might have been his favorite, she explained, "I think he mostly liked the end credits, watching the words come by and listening to all the music. That was like his favorite part. He forwarded to music and the end credits. That was a lot with watching the words come by." Existing research and anecdotal evidence also suggest that many autistic children may derive sensory pleasure from the visual motion of moving words, particularly movie credits.[35]

Kevin's consumption of text was deeply tied to his creation of textual objects. Rebecca detailed how Kevin would frequently spell out words he saw in one medium through another one. For instance, words that appeared on the iPad, Kevin would write on his Magna Doodle, a portable toy with a magnetic drawing board and stylus. When I first observed Kevin, accompanying Rebecca to a Proloquo2Go training session, he spent much of the time playing on his Magna Doodle while his mother trained on the iPad. As we sat in the family's playroom, I remarked to Rebecca that Kevin's Magna Doodle on the floor beside me slightly resembled an iPad, to which she replied, "Sometimes, I'll go [to Kevin], 'So you want the iPad?' No, he wants [the Magna Doodle]. Or he's playing with that *with* the iPad because he's writing something as he's playing the game, like 'Level One,' and so many points or something."

Kevin would also recall words that appeared on the television screen as part of DVD menus, and then reassemble those words in the form of letter tiles from the boardless board game Bananagrams. In fact, Rebecca said, "He spelled 'Indiana Jones' before he could spell his own name, with the little letter titles." The *Harry Potter* DVD menu in particular supplied rich seed material. Explained Rebecca, "He would spell 'prologue.' Prologue was his word. Prologue, prologue, prologue. Then he would spell 'quidditch pitch.' He would spell 'Florean Fortescue's Ice Cream Parlor.'"

Kevin's wordplay with the language of DVD menus also provided an opportunity for further learning. "Some things he would misspell, and you would have to correct him, and he'd want it to be the other way," Rebecca observed. "Then after a while, you'd break him, and he'd spell it the correct way." While clinicians tend to pathologize repeated viewing of movie credits by autistic youth, Rebecca described these actions as a productive pathway to spelling.[36]

Kevin not only played with multiple modes of textual representation but spatial and visual representations as well. Rebecca noted how Kevin liked to play games on the family-owned iPad, particularly Minecraft (developed by Mojang) and Pixel Gun 3D (developed by Rilisoft). Kevin was drawn to these computer games because they offered players the ability to create custom avatar skins using simple editing tools. Rebecca said of the two video games, "They've got these blocky people, but you can color their head and arms and body. It's kind of like if you took the head and then opened it flat like a box, and you paint each panel to make them look different. He *loves* doing that." Rebecca emphasized how she understood Kevin's construction of the avatars in both 2-D and 3-D as evidence of his strong spatial skills. "He's really understanding that this is the front, this is the top, this is the side and back. And he does it at lightning speed," she said.

Kevin especially enjoyed creating video game avatars based on visual representations of his favorite media characters. Noted Rebecca, "He'll look at his Lego book, see the character, and make a 'pixelized' character from his book. And I was like, 'Wow, look at that! That guy is wearing a tuxedo,' or, 'This guy is Batman.'" Kevin was unable to articulate his grasp of the English language through embodied oral speech, but Rebecca indicated that he demonstrated strengths in print literacy along with an array of new media literacies including design literacy and visual literacy. Both Cory and

Kevin's mothers described how their sons' love of words was coconstituted with their love of media in various digital and nondigital forms.

Toward an Asset Model of Disabled Children's Media Use

Parents' interpretations of their children's engagement with media as valuable were intertwined with conceptions of their nonspeaking children as social, emotional, intelligent, and verbal people. The above stories challenge deficit models of disabled children's media use, both in terms of what nonspeaking children supposedly lack and how popular culture purportedly diminishes children's development. Drawing on an expressly asset model of media education, I propose a related asset model of disabled children's media use. Such a model assumes that mass media and popular culture serve as an expansion of disabled children's communicative capacities instead of a limitation.

Due to the clinical nature of the field of speech-language pathology, whereby identifying and isolating biomedical deficits in the disabled individual repairs breakdowns in communication, the ritualistic, symbolic, and material aspects of communication are often underemphasized.[37] AAC users, for example, may choose to textually and visually express themselves through social media sites such as Instagram and Pinterest.[38] I argue that children's consumption, creation, and circulation of new media and popular culture provide a significant communicative alternative and augmentation to speech production. Stephanie did not use her iPad and the Proloquo2Go app to speak or type to Marisa the sentence, "I'm going to take care of you, Mom, because you take care of me"; Marisa interpreted Stephanie's behavior (curling up with her unwell mother to watch television) as saying so. Parents found the manner in which their nonspeaking children accessed, manipulated, and interacted with media as conveying meaning because it communicated something their child could not express through embodied oral speech.

This also suggests that youth with communicative disabilities can benefit from showcasing their verbal abilities through multimodality. Emerging readers and writers make meaning out of available cultural materials.[39] These tools for meaning making increasingly include components from digital media and popular culture.[40] A small but growing body of research indicates that emerging readers and writers with physical, cognitive, and intellectual disabilities may especially benefit from expanded opportunities to draw on their experiences with popular culture as well as leverage their multimodal text-making abilities.[41] Rebecca, to cite one illustration, aptly described Kevin as "very creative. He mixes things together." Children purposefully combine and repurpose these components through "media mixes" and "transmedia play."[42] Children with significant speech impairments communicate with the materials they have on hand, which might be the iPad at their fingertips or a handful of Bananagram tiles.

The promise of an asset model approach, however, is limited by institutional biases. Structural inequality shapes how doctors, therapists, and educators do or do not perceive parents' accounts to be relevant or credible. Working- and lower-class parents as well as those without a college degree had a more difficult time parlaying their children's engagement with media and technology at home into improved conditions and opportunities at school. This is likely due to multiple layers of disconnect between home and school culture. Parents like Marisa and Kameelah explained how they were exhausted by having to constantly convince experts that their nonspeaking child could relate to others, feel, and think. Further research is needed to understand the extent to which cultural and class biases impact institutional perception of the value of individual disabled children's out-of-school media use.

Conclusion

In sum, recreational media and technology use can help nonspeaking children reveal a side of themselves that the scientific, medical, and educational communities either do not or choose not to acknowledge. The asset model proposed here of disabled children's engagement with media rejects both a deficit model of disability and deficit model of children's popular culture. Embracing an asset model widens opportunities for children to communicate, with or without speech. This view enables us to imagine a world with greater collective communicative power, for it extends recognition of competence that is often not presumed among children and individuals with communication disabilities.

The specter of a dark and not-so-distant past hangs over this discussion—a past in which institutionalized children without the capacity for embodied oral speech spent their days parked in front of television sets with no means to communicate their needs, preferences, or the desire to change the channel. In the prologue to his memoir, *Ghost Boy: The Miraculous Escape of a Misdiagnosed Boy Trapped Inside His Own Body*, author and AAC user Martin Pistorius recalls such a childhood in a South African day care center in the 1990s:

> Barney the Dinosaur is on the TV again. I hate Barney—and his theme tune. It's sung to the tune of "Yankee Doodle Dandy." I watch children hop, skip, and jump into the huge purple dinosaur's open arms before I look around me at the room. The children here lie motionless on the floor or slumped in seats. A strap holds me upright in my wheelchair. My body, like theirs, is a prison that I can't escape: when I try to speak, I'm silent; when I will my arm to move, it stays still. There is just one difference between me and these children: my mind leaps and swoops, turns cartwheels, and somersaults as it tries to break free of its confines, conjuring a lightning flash of glorious color in a world of gray. But no one knows because I can't tell them. They think I'm an empty shell, which is why I've been sitting here listening to *Barney* or *The Lion King* day in, day out for the past nine years, and just when I thought it couldn't get any worse, *Teletubbies* came along.[43]

When he developed locked-in syndrome as an adolescent, Pistorius's parents were told by clinicians that he had the mental capacity of a three-month-old. The staff at his care home planned his media consumption accordingly. When he was twenty-five, though, a care worker noticed Pistorius's responsiveness to specific statements and, at her urging, his parents had him reevaluated. It took years for Pistorius to be able to operate an AAC device independently and put down the story above. While technology critics warn of mobile devices impairing our ability to be alone with our thoughts, glorifying solitude and a device-free life, Pistorius's story provides a powerful counterargument: being denied communication technologies can be its own form of solitary confinement.[44]

Pistorius writes of his awareness of the long journey ahead after his initial assessment by speech therapists, "This will not be a Hollywood movie with a neat happy ending or a trip to Lourdes where the mute are miraculously given a voice."[45] The next chapter interrogates the powerful media representations and narratives about voice to which Pistorius alludes. It delves deeper into the cultural spaces in which more and less privileged parents make sense of their child's disability and assistive technology use through their own engagement with various media—print, mass, independent, grassroots, and social—finding that while mediated spaces in the twenty-first century are growing more inclusive and accepting of disability, they remain exclusive to those parents with more distinctive forms of social, cultural, and economic capital.

6 "You've Gotta Be Plugged In": How Do Media Shape Understandings of the iPad?

Meryl: I had one other parent call [the iPad] "magic."

Thomas's mom, Daisy: For me—"magic"—I can say that. Because, see? Now [Thomas] can tell us. Even though we've had this since the beginning of the school year, we're still finding a lot of the "wow moments." ... You know that there's something there. But yeah, thank Steve Jobs for that! Can you imagine?

Meryl: Yeah, that's thing. I guess you don't want to use the word "revolutionary."

Daisy: I think it *is*. 'Cause there's the touch screen before. But just to do that, and then the software developers—that's *amazing*.

Daisy had to correct me not once but rather twice during our interview. She *did* want to use the words "magic" and "revolutionary" to describe the iPad. While Daisy recognized that touch screen computers existed long before the debut of iPad, when combined with Proloquo2Go, the technology was "*amazing*." Additionally, she *did* want to single out Jobs as responsible for the iPad and its continual benefits to her son's communication. While I introduced the term "revolutionary" into the conversation, my academic training had instilled in me a kind of skepticism regarding such characterizations, wary of techno-utopianism, romantic individualism, and cult of personality around industry figures like Jobs.[1]

I falsely assumed that Daisy, college educated and upper middle class, was similarly reluctant; instead, I had projected my tastes onto her. "I can say that," she told me, claiming as her own the language that I was implicitly placing judgment on. She had experienced firsthand what she called "wow moments," and was continuing to experience them. These moments provided glimpses into "something there," something to suggest that Thomas was comprehending the world around him. In all, Daisy asserted that she was licensed to speak about the iPad, Jobs, and developers of Proloquo2Go in a way that I was not.

It was a pivotal moment in my research, and one that required me to reflect on the boundary that Daisy drew between my language and hers. Before his death, Jobs, speaking on behalf of Apple, deflected any credit for the benefits that children with communication difficulties derived from the iPad. As the *Wall Street Journal* remarked in an interview with Jobs, "'We take no credit for this, and that's not our intention,' Mr. Jobs said, adding that the emails he gets from parents resonate with him."[2] A generous explanation for this quote would be that Jobs spoke out of humility, and a likely interpretation would be that his statement was carefully worded at the advice of Apple's legal team in order to distance the company from any medical liability. As Ryan Budish notes, "Certain features are unlikely to ever exist in a consumer wearable, unless [Apple CEO] Tim Cook wants to sell watches that require a doctor's prescription."[3]

Jobs's disavowal also suggests that Apple's marketing team likely prefers to associate its brand with wellness, health care, and diagnosis (e.g., its Health app, native to all Apple devices in iOS 8, and HealthKit software for app developers) rather than sickness, illness, and impairment. Tellingly, AAC device companies such as Dynavox and Tobii contributed public comments to Medicare's proposed policy changes to medical insurance coverage for speech-generating devices, while Apple stayed silent.[4]

In the *Wall Street Journal* interview, Jobs continued, "'Our intention is to say something is going on here,' and researchers should 'take a look at this.'"[5] As a communication researcher, I also consider Jobs's public statements ripe for analysis. Little is known about how parents of children who use AAC devices interpret such media messages about Apple devices and mobile technologies.[6] If I was to understand what the iPad meant to parents like Daisy and how the technology was incorporated into families' lives, then I also needed to pay closer attention to what they thought about its cultural representations.

The ways in which parents privately talk about their experiences with technology often reflect public discourses about home computing. Parents reconcile these ideas with their own realities and discussions with relatives, friends, and coworkers.[7] For example, Elaine Lally writes about how parents in the 1990s making sense of the personal computer at home projected their feelings about global technology companies (which are otherwise abstract and alienating entities) onto figures like Bill Gates.[8] Beyond the technology sector, this myth of the sole visionary plays a role in public perceptions of leaders in the creative media industries.[9]

In addition to media reports explicitly about Apple, the iPad, and Proloquo2Go, I found that parents called on, shared, and contributed content to a broad cultural imaginary concerning disability, parenting, and assistive technologies. Anthropologist Faye Ginsburg calls this space the "disability media world."[10] Rayna Rapp and Ginsburg describe this landscape as "ranging from books to documentary work to disability film festivals and screening series to YouTube uploads."[11] Many of the parents I spoke with participated in the consumption, circulation, and creation of media about parenting a child who communicates through AAC. Some of this content was specifically related to autism, which was partly due to the composition of the study's participants, autism rates in the United States, and the prevalence of media reports on autism and AAC.[12]

In the pages that follow, I detail how parents of varying class and cultural backgrounds engaged in information seeking, participated in online communities, and viewed media representations of individuals with communication disabilities or other marked forms of difference as a means of inspiration, site of resistance, and lingua franca to voice frustrations. I argue that the potential of disability media worlds to radically reconfigure family life and systems of social support in the United States is limited without also taking intersectionality into account as well as critically examining the politics of voice and recognition.

Rendering Parenting through Disability Media Worlds

I first turn to the existing literature on how parents of youth with disabilities negotiate their membership in disability media worlds. Anthropologists use the term "media worlds" to recognize the integral role that media play in everyday life as well as the multiple social and cultural fields within which media practices are situated.[13] Mass consumption of popular commodities does not make all consumers the same; rather, consumers make popular commodities into something that matches their unique lives.[14] Marginalized populations in particular use media as a resource to reimagine society and their place in it.[15] Rapp and Ginsburg contend that the stories that individuals with disabilities and their family members tell through media worlds enable the writing of alternative cultural scripts along with the revision of dominant narratives about disability.[16] These media worlds, they contend, "are crucial for building a social fund of knowledge more inclusive of the fact of disability."[17]

While disability media worlds enable the circulation of positive representations of disability and critiques of normative US family life, they also have a history of being for individuals who in other ways identify with majority culture.[18] Print media addressing individuals with disabilities and/or their parents often assume a white, highly educated, and upper-middle-class audience.[19] For example, US citizens with significant physical impairments in the post–World War II era made and distributed disability-community periodicals and newsletters.[20] The publishers, writers, and readers of these publications were relatively privileged, pointing to more systemic issues historically structuring participation in spaces for reimagining disability.

Among parents of youth with disabilities, it is unclear how class and other dimensions of difference intersect with their participation or nonparticipation in disability media worlds, or if their participation manifests in less immediately visible ways. Studies of social media and Internet use among parents of youth with disabilities frequently do not include data on socioeconomic status or racial and ethnic composition.[21] Some admittedly oversample parents with graduate degrees.[22] One content analysis of blogs maintained by "cybermothers" of children on the autism spectrum notes that the study is limited to the extent that it represents "the expression of a subgroup of parents with access [to] and proficiency in technology sufficient to establish and maintain public blogs."[23]

Issues of unequal technological "access" and "proficiency" should not be taken lightly. I found that parents primarily took part in the disability media world in five ways, with significant variations in participation by social class. First, the process of seeking out media concerning the iPad and Proloquo2Go varied between more and less privileged parents, and even among privileged parents with professional expertise in special education and assistive technology. Second, for those parents who participated in online communities concerning their child's disability or their AAC use, social media simultaneously enabled social and emotional support for some while contributing to a sense of alienation for others. Parents spoke, too, of drawing hope from viewing or interacting with ordinary AAC users in the mass media or on social media. Fourth, they resisted comparisons to high-profile AAC users whose experiences they felt were unrepresentative of their child's lives. Fifth, less privileged parents who felt especially marginalized in their interactions with educational and clinical professionals drew on media to explain their experiences of isolation and being misunderstood. These distinctions combined have significant implications for the future inclusivity of disability media worlds and the potential for parents to overcome power imbalances in social structures.

Information Seeking

A clear difference in information-seeking behaviors emerged between more and less privileged parents. The former tended to be highly educated and tech savvy, and the latter less formally educated, proficient in English, and comfortable with using technology. The first group regularly sought out media to learn new ways to help support their children's communication, particularly by conducting Internet searches, while the second preferred to support their children by gaining insights through lived experience, scaffolded learning, and analog materials. Tensions arose when well-meaning professionals, belonging to the same cultural milieu as more privileged parents, recommended the Internet as a primary information resource to less privileged parents. The Internet may be a cost-effective and widely distributed approach to supplying caregivers with information, but in light of corresponding budgetary restrictions on social services, this solution can reinforce class stratification by requiring all parents to mobilize unequally distributed social and cultural capital.

Always Looking for Answers

Chike's mom, Esosa, mentioned that she had seen a piece about iPads as AAC devices on *60 Minutes*. She was referring to a segment titled "Apps for Autism" that originally aired in the United States on October 23, 2011 (and was later made available online through CBS News' YouTube page). I asked Esosa if she thought that television reports such as this one influenced parents in any way. "Yeah," said Esosa, "because we're always looking for, you know. ... If you see something about autism or read about it, you pick out what is going to apply to you. If it doesn't apply to you, or you've tried it, you discount it." Esosa counted herself among parents who are always looking for novel and relevant information to help their nonspeaking child, including information about the latest technology. One assistive technology school district administrator remarked that it was the "parent who is always looking for the answer of what's going to happen" who would be most likely to make the investment of buying an iPad and AAC app.

In an era when online information about disability causes, diagnoses, and treatments can be highly misleading as well as patently false (for instance, the debunked linkage between vaccines and autism), judging the accuracy of media reports about the role of new media in the lives of disabled youth requires a level of literacy and discernment.[24] Take Peter, who reported that "every once in a while, while I've got five minutes at work, I'll Google 'new toddler apps' and just look at them." Instead of searching for "apps for autism," he had developed a better search strategy: "[I look] for quality kid apps, and then I read the review and look at it and say maybe it'll work, maybe it won't." Peter also relied on recommendations from autism forums on Facebook: "Somebody will always say, 'My kid just got an iPad. What apps?' Usually, I'm like 'Got it. Got it. Ooh! Haven't heard of that one!' Facebook has been great." Daisy noted, "I just Google 'iPad autism.' ... I just Google all the time. OK, what's out there, what's going on?"

When Isaac's teacher suggested that he use a visual schedule app on the iPod to aid his integration in mainstream kindergarten, Sara too turned to the Internet. "I went online, I did my research," she remarked. Their responses suggest an ongoing and dynamic process of more privileged parents "doing their research" by navigating online media and interpreting content as situationally relevant for their child. Because they are frequently seeking out information across multiple platforms, more privileged parents also perceive there to be a constant flow of news about nonspeaking children and technology. Sara, for instance, said she had seen "all these programs on TV about these kids who were doing all these crazy things, and look, this teacher found [Proloquo2Go]."

A number of the assistive technology specialists whom I interviewed mentioned such depictions having an impact on upper-middle-class parents, leading some to insist that their child be evaluated for an iPad. An AAC coordinator, one who manages a caseload in a privileged neighborhood within a public school district otherwise underresourced, recalled a day when roughly ten speech-language pathologists in schools she worked with had parents requesting, "'I want an iPad for my kid. I want them to talk.' Then you're like, 'Oh, where did you hear this from?'" The coordinator asked her coworkers, "'Do you know if there were any interviews or anything on TV?' Sure enough, there was something that had been replayed. It was an old interview on an old show." The news cycle propagates stories about the miraculous benefits of technology—stories that more privileged parents are increasingly likely to be on the lookout for and find. These same parents also mobilize their social and cultural capital to request that schools provide more assistive technology support for their child. "I don't blame parents at all," said the coordinator. "They just want an answer."

Always Surrounded by Firsthand Information

A couple of highly educated parents with whom I spoke also worked in the special education or assistive technology field. These parents were not so much always *looking* for information about assistive technology and AAC devices as they were quite atypically always *surrounded* by it. When I mentioned to Nina, a special education professor, that I was studying Proloquo2Go because it was the dominant player in the AAC app field, she agreed, saying, "The media has highlighted it *so* much." When she wanted to research how to model use of the Proloquo2Go for her son, Nina said, "I started looking on Facebook. I'm friends with AAC people [laughs], and they were like, 'Modeling, modeling, modeling,' and aided language stimulation and all this stuff." Through these connections, Nina found two other helpful websites: the blog *PrAACtical AAC* and the AAC Language Lab website. "I've used those to help me, and that's how I learned about modeling," Nina explained.

Donna thought it was "amazing that these parents will see it on maybe *20/20* or they'll read it in the magazine," but she also recognized her own privileged position due to her part-time job at a disability resource center. She noted, "I actually got to see this stuff firsthand, and actually utilize it and see it. I think I was extremely lucky to be in the right place at the right time." In turn, Donna considered herself a human resource to parents in her community. "Nobody's going to tell them unless they see it on *20/20* or they see me at the grocery store," she said, referring to her son's AAC device. "I've been stopped, I can't tell you how many times, with grandmas, teachers, just wanting to know more about it to perhaps help out a student or a family member."

Not Hungry for Knowledge

In their exploratory study of how parents of autistic youth seek information online, Jennifer Reinke and Catherine Solheim identified a main theme—"seeking is an ongoing part of my life"—that aptly describes the more privileged parents discussed above.[25] Similarly, Tawfiq Ammari and Sarita Schoenebeck found that highly educated parents of youth with disabilities on Facebook use the site as a way to gather experiential information.[26] The patterns identified in these studies cannot be generalized, though, to all parents of youth on the autism spectrum or youth with communication disabilities. For example, there may be significant differences between parents of higher and lower socioeconomic status in their use of the Internet to obtain information about caring for their child with a disability, with the cost of computer equipment and Internet access serving as a financial barrier for low-income parents.[27]

If I had only spoken to more privileged parents who were either always looking for or always surrounded by information through their social networks, I might have overlooked all the other parents of nonspeaking children who had amassed less distinctive forms of social, cultural, and economic capital. Less privileged parents—many of whom did not have a college degree, speak English as a native language, or consider themselves "tech savvy"—were not as likely as more privileged parents to seek out mass media or turn to English-dominated spaces on the Internet for information about how technology might support their child. They also made fewer direct demands of school districts to provide technological solutions based on having encountered a relevant story in the media.

Yet it would be a mistake to assume that just because these parents did not search for information in the same manner as more privileged parents that they cared any less about their children. One AAC consultant highlighted the structural constraints on less privileged parents' time for information seeking, as compared to more privileged individuals:

> There are some gung ho parents who will do anything. They try. They learn. They do YouTube videos on how. With the Internet, anything's accessible online. They train themselves. But it all comes down to ... and do they have five jobs? Do they even have time to train themselves? A lot of these families, you know, parents have two or three jobs. They don't have time. It comes down to, is this a stay-at-home mom who has time to just dedicate five hours a day to do training? It's sad. I don't think that parents aren't motivated. I think it just comes down to circumstances.

Given their position, I found that less privileged parents instead preferred to learn how to better support their child through hands-on experiences, print materials, and culturally appropriate programs. David, for instance, described feeling frustrated that he could not do more for Beatriz. "I always tell everyone—her doctor, her teacher. I always tell them that my main concern and my main interest is to learn how to communicate with her," said David. "I'm always trying to find ways, but ... I'm limited because, you know, what can I do? I just read."

In response to his mention of reading, I asked, "Where do you find information? Do you read it online?" "I go online," he replied, "but I don't ... you think I'm ... ?" I had made a misstep by expecting David to turn to the Internet for answers, likely because of my own class and educational background (as I had with Daisy). "Yeah, I read," he continued, "but not like, 'Oh, I'll go here, and then I'm hungry for knowledge about this type of situation.' No." Whereas Daisy used Google frequently and flexibly, David drew more on tacit knowledge and less on information circulating in the media in order to help his child. "When I talk to you," he said to me, "I'm just talking about my experience with this device, and the program, and the previous program that she had."

Rebecca admitted being afraid to touch the computer, saying, "If I do something, it's going to disappear forever. I'm afraid of permanently screwing something up. I'm reluctant to just tap away." None of the more privileged mothers I talked to expressed this level of reluctance. Rebecca recalled being "flummoxed" by a recent "registration for my son online" because the print instructions she had been given did not match what appeared on the screen. She was unsatisfied with the response given by those in charge of the website. "It was like, 'I think you just have to ignore that. Just answer this.' Just have it be like, this *does* refer to that," Rebecca protested. "If they're going to keep changing stuff, they need to have reference material to keep up with that."

When it came to learning how to use Proloquo2Go, Rebecca disclosed that her lack of comfort with technology created a barrier between her and online resources. "I don't like having to look it up online to see how to do it because if you're not 'technologically savvy' [gestures with air quotes], you don't know how to look for things online, or maybe you don't have access to things online." Instead, she deferred to the expertise of her husband. "He's really good on the computer because that's what he does," said Rebecca. "His design work is on the computer."

Rather than being required to use the Internet, Rebecca preferred learning through free, hands-on Proloquo2Go workshops (where I first met her) and watching another person demonstrate. She described the ideal way to learn about Proloquo2Go, drawing an analogy to learning a craft. "I do knitting and crochet," Rebecca explained. "Certain things, you need to see somebody doing it. Seeing it static in a book sometimes doesn't help on certain things. Like, 'You need to swipe this way.'" While Rebecca had not attended college, she had spent time at an art school after high school. An online instructional video might best serve a tech-savvy parent, but the Internet skills needed to access that video created a barrier for Rebecca.

Similarly, Vanessa did not complete college and picked up on the technological expertise of men in her life. She felt more capable handling computers when led through a live demonstration and she could take handwritten notes. Vanessa explained,

> Fortunately, with [Moira's] father and my boyfriend, with Brandon, they're very tech savvy, and I'm not. I'm just like, "Hey. Can you show me how to do this?" Both of them have always been like, "Yeah. This is how you do it." Then I write it down, and I'm like, "All right, cool!" and then I pick it up. I'm not tech savvy. If someone explains it to me, and then if I write it down and I have it for reference, then I'm good.

Vanessa placed the blame on herself when having difficulty in navigating the iPad, internalizing the expectation that because the technology was so "user friendly," any problems were due to her own overthinking. Vanessa said, "There have been times where I'm like, 'How do you do this?' 'I wish I could just do this,' and then it does it. ... They make [the iPad] so user friendly, that you make it more complicated than it actually is."

Beyond computer-based resources, less privileged parents found libraries and parent support groups in their local communities helpful for learning more about assistive technologies for communication. The library was the first stop for Stephanie's mom, Marisa, when her daughter was diagnosed with autism. Marisa commented that she didn't feel comfortable with technology, and that even their home computer printer was "too much technology." When Stephanie was diagnosed, "I don't know how to teach my daughter to speak," Marisa said. "I went to the library and get sign language books. I start to learn the basic like, 'please,' 'thank you,' 'more.' I start to teach Stephanie, but was not enough."

Marisa and Nelson sought the wisdom of others nearby to better equip themselves. They joined a community support group for parents of autistic children, which was where they first learned about Proloquo2Go. Said Nelson, "They give special seminars, and different public speakers come out. This particular time, there was one on the iPad." Karun, who was involved in her local Armenian church and cultural organization, mentioned that a support group for Armenian parents of autistic children "helped me a lot." Through the group, she was able to connect with other parents who shared her cultural heritage. Less privileged parents were not "hungry for knowledge," to use David's terminology, in the sense that they consumed mass amounts of information from the Internet, but instead actively sought out alternative channels that best fit their lives and backgrounds.

More privileged parents also mentioned local groups as being helpful resources for information about using the iPad for AAC, but tended to mention such organizations in the context of their online presence. Alice and Peter said that their local regional center's volunteer arm, called Coastal Coalition, had an active Facebook page. Without a Facebook account, though, one would be disconnected from the groups' in-person events. "A lot of times you have to make sure that you're 'liking' these people on Facebook or that you're somehow plugged into them some other way," explained Alice. "You have to initiate and after that you get stuff." Peter concurred, noting, "I do wish there was a clearinghouse for data or information. There is not. ... You've gotta be plugged in." Being plugged in involves more than just technical access to the Internet and parents' independent initiative; it requires access to social and cultural capital, which is bound to infrastructures and institutions.

Online Video Tutorials

Between these three groups of parents—the always looking, always surrounded, and not hungry—it becomes clear that the Internet is not a one-size-fits-all solution for informing parents about how they might best support their children's communication through assistive technology. These distinctions were particularly evident when it came to whether or not parents watched online video tutorials for Proloquo2Go. Tech-savvy parents tended to find these helpful. Anne, for one, said, "We watched when we were researching [Proloquo2Go]. We watched a couple of YouTube videos." Rachel and Caren, the assistive technology specialists whom I observed, frequently recommended these tutorials and the AssistiveWare website to parents. Less privileged parents, however, were not as receptive to being directed to the Internet. When I asked David if he had visited the Proloquo2Go website, he replied, "I'm gonna be honest with you, I haven't. And Rachel gave me a website, and she told me there is a lot of resources on their website, but I haven't."

Even though David had reliable Internet access at home, using the web to watch tutorial videos was a relatively low priority. Similarly, when Rachel asked Karun if she'd been watching how-to videos on the Proloquo2Go site, Karun responded no. "I've been busy with [Pargev's] IEP," she said, and trying to start her small business. Something that came up in my later interview with Karun made me recall this exchange with Rachel. During the interview, Karun told me about how she first heard about Proloquo2Go from Pargev's teacher, shortly after fleeing Syria and seeking asylum in California. The teacher "kept on telling me you have to download it. You have to buy iPad," said Karun. "But we couldn't buy it because we came from the war with no money." Trained school professionals were understandably, but frustratingly, not fully in touch with the realities of working-class households.

Teachers and therapists were disappointed in less privileged parents for seemingly slacking, not fully comprehending that their cultural values, habits, and preferences around information seeking were different from their own. After Marisa admitted to Rachel that she wasn't comfortable navigating Proloquo2Go on the iPad, Rachel did not seem to completely grasp Marisa's reticence. Said Rachel, a bit dismissively, "You have an iPhone, so you can do this." Rachel then used a pen and spare sheet of paper to write down the URL for the AssistiveWare website, suggesting that Stephanie's parents turn there for "information and webinars," and then "e-mail any questions" to the company directly, assuming that parents would be comfortable engaging in such a practice. Marisa was unsatisfied with this offer, and asked Rachel if she could possibly visit their home again to work one-on-one with the family. With a tightened disability resource center budget and limited paid hours allotted for at-home technology training sessions, Rachel said that was unlikely. A website was all that she had to offer—an insufficient resource for families that needed her highly skilled support the most.

Participation in Online Communities

Information and social support tend to be the primary motives for parents to engage in online social networks such as Facebook.[28] While less privileged parents reported being less inclined than more privileged ones to seek out information on the Internet, both found that online communities provided opportunities for social and emotional support. Vanessa noted, "I'm friends with quite a few moms on Facebook whom I've never met in actual life, but we've become close over the years and support each other." Some women discussed how online communities offered spaces for alternative conceptions of mothering a child with a disability; others found that they also reinforced unrealistic expectations. While prior work has focused primarily on the organization of online communities for parents of children with disabilities by geography or the specific disability, parents also spoke of the pros and cons of online groups organized based on specific AAC platforms.[29]

Social and Emotional Support

Perri gave one of the most passionate endorsements for Facebook that I have ever heard. Referring to a Facebook group for parents of children with one of her child's multiple rare disorders, she said, "I mean that's what Facebook was meant to be. Really. I'm not a big Facebook user, but that has been instrumental in our lives to connect with the world of children with [Cory's disorder]. It's been amazing." Parents of children with rare conditions are especially likely to take part in online communities by consuming, sharing, and circulating content.[30]

Sometimes, the sharing of information about Proloquo2Go as well as supplying social and emotional support on Facebook were inherently entwined. Perri explained,

> Cory has like a little twin guy who lives in Missouri who looks very much like him. I guess [he] is one of three children who has his genetic makeup of the way [that rare disorder has affected Cory]. I saw that [his mom] posted last night something about Proloquo, so I'm very interested in talking to [her] about it as well and have to see how Parker is using it. You know, how it's going with him. I was sad to see him in a wheelchair because I know he was walking. I have to figure that out. Understand that more.

Facebook and other social media may provide a particularly useful outlet for geographically dispersed parents of children with rare conditions to build meaningful relationships.

Researchers have also noted that parents of children with disabilities engage with other parents both through their profile pages and private groups on Facebook.[31] Beyond these components, Facebook's messaging function serves parents in developing intimate one-on-one relationships for social support. Vanessa discussed developing a friendship with another parent independent of a formal group, but cultivated through a preexisting social connection. She explained, "There is a mom whom I've only met once who lives in Cheyenne, [Wyoming] and she's friends with one of my childhood friends, and her son and Moira have very similar challenges." In a more private corner of the social media site, "I'm able to message her on Facebook and really get down to how it's affecting me, and she totally gets it," confided Vanessa. Parents build intimacy across public forms, semiprivate groups, and private messaging on Facebook.

Having moved recently, Vanessa and Moira had yet to establish themselves in their community. Most of the moms whom Vanessa was currently meeting locally did not "get" her. She said, "Have *not* connected with moms in [Moira's] classrooms, which is a bit boggling. It could just be because we are so overextended and overwhelmed." In comparison, within their old town, "I definitely had a network of moms whom I've connected with over the years through Moira's services, meeting moms in the waiting rooms." Vanessa disliked the support groups she encountered in her new city because they were "all about strategies, networking, schools, and special education issues, and special education law, IEPs and various therapies. ... [I]t's always strategizing." She had yet to find a support group "where I can really be able to express my fears and anxieties," and "that really talks about *us*." Vanessa opened up to me about these feelings:

> I have been starting to go through a thing lately where I'm feeling a bit resentful of how much this has consumed my life and I feel I'm having to try to reestablish where *my* identity is in this because literally, my entire life is consumed with her

> care. When she's at school, I'm on the phone with regional center and school, and texting with her teacher, scheduling her appointments and taking her to appointments, meeting with the behaviorists, working on her programs with the behaviorists, and then developing my own things that I'm trying to work with her on. I stopped working. I couldn't manage working and her needs.

Vanessa was in no way resentful of her daughter; she longed, though, for a group that let her explore her identity outside the role of Moira's personal special education strategist. Her social ties to other mothers over Facebook were her most secure ones, as Facebook provided her with the therapeutic space that she did not find among the parent support groups in her new geographic community.

Social Comparison

Vanessa sought to connect with other mothers in ways that could help release the pressures of the intensive mothering/warrior-hero paradigm. Intensive mothering is a contemporary child-rearing ideology in which "good mothers" are expected to self-sacrifice and prioritize the needs of their children above their own personal as well as professional interests.[32] Mothers (especially white, upper-middle-class ones) of US children with disabilities are culturally expected to be on constant guard against violations of their child's educational rights.[33] A subcategory of the good mother archetype is that of the "warrior-hero." Such mothers, Amy Sousa writes, wage "battle against social and political forces to gain medical and educational interventions for their children despite the high personal and financial costs to themselves and their families."[34] Esosa described encountering such a mother online:

> There's one on that Facebook group that we belong to. Somebody posted, it's this mom whose son was autistic, and the doctors basically said, "This is how it's going to be. Forget about it." She noticed that he liked looking at glass. She used to surround him with glasses and she would fill them with water. Now he's studying and he's in college doing astrophysics or something like that.

Not only did this mother resist the doctors' classification of her son, according to Esosa, but she helped him to supersede academic expectations of typically developing children as well. This encounter on Facebook was an opportunity for Esosa to reflect on her own parenting efforts. "When I read that," said Esosa, "I thought, instead of fighting [Chike's] stimming, I should figure out how to, like this mom did, she kind of directed it somehow and it turned out positive."

Yet Esosa also recognized the inherent challenges in actually achieving that goal and the futility of social comparison. While Esosa had a live-at-home nanny from Mexico and her husband to assist her, she also worked full-time and was raising four children in total. "I don't know how many other kids she had," Esosa said, referring to the mother in her Facebook group, but "really, if you're going to be successful with a child like this, you have to drop everything and give up to be the center of your life, which you can't really do." Not only was this ideal mother devoted around the clock, explained Esosa, but she would also have to be creative, flexible, and patient.

It should be noted that these social comparisons with other mothers were not exclusive to online spaces. When discussing ways she learned new things about assistive technology, Rebecca brought up a mother of another nonspeaking autistic child in Kevin's after-school peer group. As part of a learning program she had developed herself, the mother was assisting her son in using the iPad to communicate (not through Proloquo2Go). "She's got a little wireless keyboard and she types out a question," said Rebecca. "That's opened up the door for her son to be more interactive with the peers in our group." Rebecca admired the other mother for being "very hands on," while recognizing that the woman's success was partly due to her accumulated cultural capital. Rebecca remarked, "She's also got a doctor background so she's very sophisticated, very educated. Her husband's a heart specialist. They have a very educated background, and so she's able to look at things in a different way and maybe be more pragmatic." Rebecca recognized that upper-middle-class mothers, especially those with medical backgrounds, tend to have more resources and interest, as Colin Ong-Dean writes, "not only in improving their children's knowledge and skills but also in intervening in the institutional processes for evaluating their children's abilities and accommodating their needs."[35]

Device-Based Groups

In their study of how parents of children with disabilities use social media for information gathering and social support, Ammari, Merrie Morris, and Schoenebeck classified the social media sites parents tended to visit into two groups: geographically based (grouping parents according to location) and care based (grouping parents according to their child's condition).[36] A number of parents in my research, though, also mentioned joining official Facebook groups for parents whose children used Proloquo2Go. Nina shared, "There's a Proloquo2Go for parents group on Facebook, and that was very cool because people take screen shots. ... I remember in one of my workshops that I gave, I saved some of those screenshots." These device-based groups, like those for parents of Proloquo2Go users, are distinct from geographically and care-based groups, as parents may be geographically dispersed or caring for children with rare conditions impacting their ability to produce oral speech.

It seemed to me that Anne could have benefited from belonging to a device-based group. She mentioned feeling anxious about transitioning her son from using American Sign Language as his primary form of AAC to the iPad and Proloquo2Go. Her unease partly stemmed from feeling as though she had no other parents to identify with, besides her husband. "We've never met anybody else in exactly the same position, who isn't deaf but was raised using sign language, but then had to transition [to using an AAC device instead of sign language]," explained Anne. In her local school community, Anne could not talk about communication aids with other parents because her son's classmates were Deaf and did not use a high-tech AAC device to express themselves.

Anne was in a double bind, as she was also dissatisfied with the current care-based Facebook group to which she belonged. While "there's a Facebook group of kids with [rare disorder]," she said it "just isn't that helpful" because none of the children also used AAC. "Most of the kids, we've talked to their parents, they can talk, so it's not a thing. They have other issues," she commented. Anne made a comparison to parents whose child has a more common developmental disability, and discussed their sense of group membership. "It's just so hard," she shared, "because his thing isn't Down syndrome or something, where there's this huge community of people." While the Facebook group for parents of children with her son's rare disorder provided some form of belonging, Anne was simultaneously an outsider from it due to her son's communication needs.

Device-based groups, however, can also have their drawbacks. While Nina appreciated the utility of the Proloquo2Go for parents Facebook group as an information exchange, Peter found it and other existing device-based groups to inherently prioritize the corporate interests of the company over that of parents. He noted, "I do wish that each app, like Words for Life and Proloquo, would have a user group that was a little less commercial, because it's always like, 'Go on our site,' but they obviously edit and everything like that." Instead, he hoped for a grassroots device-based group that "was a little more free-flowing forum so parents could say, by the way, don't forget you can do this."

Even less commercial device-based online groups presented problems. Some parents perceived that in order to take part, a base level of technological literacy with AAC devices was required. Explained Perri,

> My friend, Kiara, has a son, Oskar, who has [a rare syndrome]. She has added me to the Yahoo group of the communication device thing for [this rare syndrome], and I haven't spent enough time on there. I'm trying to get over the hump myself because I know it doesn't matter whatever feedback I have from other parents: I still have to learn it, and he also has to learn it. But once we're in the groove, I'm going to, definitely.

While these device-based communities (which can also overlap with geographically and care-based groups) might be imagined as a resource for helping parents like Perri "get over the hump," novices may feel discouraged from participating, even peripherally.

Information and Inspiration from Ordinary People in the Media

Across socioeconomic status, parents discussed the value of viewing nonfictional media representations of ordinary people who use AAC systems. These individuals fell into two groups: "real" users around the child's age, and "role model" teen and adult users. Depictions featuring these people supplied parents with useful advice as well as a sense of hope. Personal experiences and anecdotes can be particularly compelling for parents, even more so than empirical evidence.[37] "Those YouTube videos, the documentaries," said Nina, "that's now where my frame of reference is. Those are my sources of information and inspiration at the same time." These depictions nevertheless may also have negative consequences, as they potentially give parents (especially less privileged ones) a sense of false hope by providing little transparency about the economic, social, and cultural resources behind such success stories.

Real Children

Many parents found it beneficial to observe "real" children using the iPad and Proloquo2Go, filmed by professionals or their own parents.[38] In fact, Daisy mentioned that viewing parent-posted videos of children using the iPad and Proloquo2Go was the main reason she visited YouTube. "You go to YouTube, and they actually show you how they're using this," she said of such home movies, "because these are real therapies happening. I guess the parents are recording it. ... Parents are excited because their kid is talking for the first time." Anne discovered Proloquo2Go in an online search for an

AAC system for the iPad that seemed age appropriate for her son. She recalled, "Basically, 'iPad speaking programs' is what we Googled, and then there were a couple of them, and the Proloquo seemed to be the one recommended for our age of child." Her and her husband were finally sold on Proloquo2Go when they saw it on film. She explained, "Then we saw some videos of kids his age using it. We could visualize how this would actually work." Anne appreciated being able to see her child in other parents' user-generated videos as an authentic testimonial.

Nina was particularly inspired by YouTube videos that featured children communicating through an approach known as Rapid Prompting Method and a related documentary she had seen on HBO titled *A Mother's Courage: Talking Back to Autism*.[39] Nina called the movie "very eye-opening. That's what gave me hope. I saw these kids who may have been considered to have 'severe behaviors,' but they were so bright, so intelligent, playing the stock market, reading these amazing books." She wanted to be able to tap into what Raul knew yet could not communicate, and Rapid Prompting Method offered a potential solution. "I don't want to give up hope," Nina said.

Segments from news magazine programs such as *20/20* and *60 Minutes* featuring young autistic AAC users were especially memorable for a number of parents. In the middle of my interview with Stephanie's parents, her father, Nelson, asked me, "Have you seen a video called the *Most Amazing Story* about a child with autism?" One of Stephanie's therapists had shown it to him, and he in turn shared it with all of his Facebook friends. One friend—the wife of a coworker who also happened to be a behavioral therapist and mother of an autistic girl—sent him "some other videos to watch. And I was like, 'Wow, OK.'"

Enthusiastically, Nelson hopped up from the dining room table to the nearby desktop computer to show me the clip. He opened a web browser and pulled up a YouTube video of *20/20*'s story on Carly Fleishmann, an adolescent, nonspeaking autistic girl who spent much of her young life without a means of communication. The video began with the following dramatic introduction from anchor John Stossel: "Now, what may be the most remarkable coming out event we've ever seen."[40] Stossel continued, "A little girl is unable to speak a word or connect to the world around her in any way. Or so everyone thought until she turned 11, and suddenly, something remarkable happened. ABC's John McKenzie, has the story of a girl breaking out of her own body." Nelson rewatched the segment, rapt. "When I saw the video, and I think as far as for my husband also," said

Marisa, "we have hope one day Stephanie type something to say, 'I'm here. I hear you.'"

Karun had a similar experience viewing the "Apps for Autism" segment from *60 Minutes*, which a relative had recorded for her to watch. She recalled the segment featuring teenagers who communicate by typing on an iPad: "When the child showed the interest or started to type suddenly. Suddenly, one of them wasn't verbal, wasn't communicating. Suddenly this iPad helped him." That Karun repeatedly mentioned the "sudden" development of the child's communication with the iPad was in part related to her hope that Pargev would do the same. Speaking of a boy featured in the *60 Minutes* piece, she said, "It came from the child. I don't know. I don't want to wait for Pargev one day just to come up from inside."

While these stories inspired parents, they might have a negative impact too. A number of clinicians have raised concerns that media reports about nonspeaking autistic children producing sophisticated original texts through communication technologies are misleading, and can potentially heighten parental stress and guilt.[41] One AAC specialist I spoke to was concerned that parents were being misled by these news reports. "In the media, they're stressing, 'Oh, my kid has autism. I've never heard them before. I threw an iPad in front of them, and all of a sudden they're telling stories and writing books,'" she said. "Maybe one out of how many kids will that happen to?" The AAC specialist felt that parents were not getting the full story. To her, these news reports accelerated the timeline of children's progress, and obscured the sustained work of teachers, therapists, parents, and especially the children themselves that led to a dramatic difference in the child's communication.

Role Models

Besides connecting with media that spoke to their child's communication needs at present, a number of parents mentioned the importance of media that reflected the experiences of older AAC users and those on the autism spectrum. Social media, books, news reports, and documentaries featuring these role models enabled parents to feel better prepared for their child's uncertain future. Nina mentioned seeing news reports ("I don't know if it was *20/20* or CNN") a few years earlier about Proloquo2Go, and "how it's opening doors for adults and teenagers with autism who are mostly nonverbal." She was inspired and recalled thinking, "Gosh! Raul can do this. I think [Proloquo2Go] would be great for him." She also raved about a documentary that featured two nonspeaking autistic men (*Wretches and Jabberers*) and a few memoirs written by nonspeaking autistic youth (*Ido in*

Autismland, Carly's Voice, and *The Reason I Jump*). "Those are all people who are role models for my son," said Nina. "I've been focusing less on speech and more on communication." While Raul had spent significant time in therapy setting oral speech production as a goal, these public figures were helping Nina to realize that a singular emphasis on verbalization was detracting from sustained efforts to support her young son's lifelong expression.

Vanessa felt that she wasn't getting the full picture, however, through these media portrayals of supposed role models. During our interview, I noticed *Carly's Voice* on Vanessa's living room bookshelf and asked what she thought of the book. "I haven't finished that actually," she told me. "I started it, and I'm like, 'All right. This is our story.'" Yet the more Vanessa read, the more she realized that Carly's story was not typical. "I want to know, how did they get to the point of [Carly] being able to fully express her experience and perspective?" she asked. "No one has been able to answer that question for me … of, 'How do we teach [Moira] to get to that point'?"

In the meantime, Vanessa concentrated on preparing for Moira's future and independence, and their eventual separation. "Everything I do with her, every interaction, everything I want to teach her is always with her later years in mind," she said. Part of this preparation involved following Facebook pages, such as *Thinking Person's Guide to Autism*, and reading blogs written by adults on the autism spectrum. These resources had been "really awesome because … I'm always thinking about her adulthood and her impending adolescence. To be able to get those perspectives is really great."

These media encounters served an additional role in that they shaped Vanessa's political positioning toward disability. The self-advocacy of autistic adults "really helped me a lot to sort through things. It's a really interesting movement of people in the autism population that are saying like, 'I'm not damaged. I don't need to be fixed.'" She explained, "I follow a group on Facebook that's a page that is operated by three people with autism. They create all these … [T]hey're not 'memes,' if it's all words, is it? When they just do an image with the phrase? I don't know. But they always say, 'I am autistic.'" Vanessa and Nina sought out adult and teenage role models for their nonspeaking autistic children through mass, online, and social media, and in doing so, learned about themselves, the kind of parents they aspired to be, and the futures they wanted for their child and our society.

Dissociation with Famous Figures

When I describe AAC to those who are not well versed, I find that making a brief reference to Hawking or Ebert helps others understand. Yet these men do not represent all AAC device users. Hawking and Ebert are white, heterosexual, cisgender, English-speaking, college-educated men. Their life stories follow similar plots: after attaining great professional success, they lost their speech due to a disability they were not born with. Their cognition was not impacted, and their status as authority figures never wavered. In fact, these men gained further veneration as poetic, inspirational icons. Although they use alternative means of communication, audiences around the world revere their words.

Parents made reference to these individuals not to draw parallels to their children but rather to contrast these men's experiences with those of their child. "I think that Stephen Hawking uses the kind of thing like Sam uses, that type of device," said Sam's mom, Donna. Despite this commonality, she emphasized that Hawking and Sam had entirely different medical conditions and cognitive abilities. While Hawking has "got the language," Sam "doesn't have language," said Donna—at least not at the moment. "[Sam's] getting it, but that's the whole thing"—"the whole thing" referring to the key distinction between her son and Hawking. "[Sam] needs to learn language from the ground up, just like a four-year-old or whatever," whereas with Hawking, "the guy's brain is fine. His body is just betraying him," she explained.

Eric's mom, Anne, also brought up a famous figure whose use of an AAC system was quite different than that of her son. She had been talking about feeling as though she had no reference point for choosing a synthetic voice that Eric could use with Proloquo2Go. Eric had never spoken before, so her and her husband chose the default boy voice on the app. In comparison, before requiring an AAC system in order to talk, Ebert had spent many decades speaking as a movie critic and cohost of the television show *At the Movies*. With "Roger Ebert, they had so many recordings of his own voice that he was able to use his own voice," said Anne, referring to Ebert's custom-made synthetic voice.[42]

Interestingly, as a way to describe Eric's particular communication needs, Anne turned away from Ebert and toward the Oscar-winning film *The Piano*.[43] The movie centers around Ada McGrath (played by Holly Hunter), who is mute, and communicates her feelings, needs, and desires through various modes, including piano playing, sign language, and a form of mobile media. Said Anne, "As [Eric] gets better at writing, I'm wondering if eventually, this is silly, but maybe like [an iPad] stylus or like ... in *The Piano*. She had that little pad of paper around her and just could write." Like Ada, Eric sometimes used a pencil and paper as part of his AAC system, deploying different mobile communication forms depending on the context. "We didn't understand him," Anne remarked. "It was just faster for him to write it down, and hand it over." While Hawking and Ebert are the most famous people to *speak through* a voice output communication aid, they do not *speak to* either Sam or Eric's experience in the eyes of their respective parents.

Mass Media and the Cultural Language of Special Education

Cultural material depicting AAC and AAC users, while far from widespread, appears sporadically in the US mass media (e.g., the television sitcom *Speechless* that debuted on ABC in fall 2016). A parent need not have obtained an advanced degree to know who Hawking is, for instance. There is, in fact, an entire Wikipedia page devoted to "Stephen Hawking in popular culture," noting his appearances as himself (*The Big Bang Theory*), in cartoon form (*The Simpsons* and *Futurama*), and portrayed by an actor (*The Theory of Everything*).[44] Less privileged parents used mass media and popular culture as alternatives to the cultural language of special education and assistive technology shared by professionals and more privileged parents, drawing on what James Carey termed a "publicly available stock of symbols."[45] Parents called on these symbols in two key ways: as a means to grasp clinical understandings of their child's disability, and to talk about their frustrations with institutional hierarchies.

Clinical Understandings

On multiple levels, Nelson's experiences with film informed his understanding of his daughter's autism. Sans college degree, Nelson had spent a number of years in Southern California doing unglamorous film production work for ten hours a day, six days a week. He currently held a part-time job as a photographer, which both generated additional income and allowed him to stay connected with his passion for film. His enthusiasm for the cinematic arts, and particularly the genre of science fiction, was integrated into his understanding of how Stephanie's autism shaped her daily life: "I look back, and I saw the *Rain Man* movie with Dustin Hoffman and *Forrest Gump*. It's an amazing movie. I saw this show called *Alphas*. ... There was a kid who was autistic, but he could see radio waves and sound waves and Internet with the spectrum. And I was like, 'Man, I wonder if my little girl sees stuff like that.'" These narratives all feature characters whose autism endows them with enhanced abilities. In fact, with wonder and appreciation, Nelson described autistic individuals as "real X-Men" based on what he had learned about a possible explanation for some of Stephanie's behaviors:

> For a long time, she would never make eye contact. Until somebody explained to me, "You're freaking her out. She just doesn't see *you*. She sees everything." I'm like, "You mean like in high definition?" And he goes, "Yeah, twenty-four frames per second." It clicked in my head. You know what? People got sick. This last movie that came out in high definition, HFR [high frame rate] or 3-, 4-D, or whatever. They have this new format that people were actually getting dizzy and nauseated, because if you see your hand in front of you, if you put it here, everything else is fuzzy back here. And if you start going out, then you focus back there, the depth of field changes. I don't know if she sees everything the same, you know? Like, if you see that in too much high definition, your brain is … it's being overwhelmed, overstimulated. They do the spins, they jump. They're trying to sooth, though. Wow. It's a trip. It's a real trip.

No explanation for Stephanie's unique way of processing sensory input "clicked" for Nelson until he thought about it in terms of metaphors with which he was already familiar—film, movie cameras, and audiences.

Voicing Frustration

Nelson's wife, Marisa, spoke about how the *20/20* piece on Fleishmann made her feel connected to a broader community of parents. "When I saw that one," she said, "it's like OK, I'm not the only one." Marisa explained the root of her isolation. "Because sometime we feel alone. Because I saw that—the district, they're professional. And they have masters. They have doctors' degrees, but they don't have a kid like Stephanie. They miss day by day." While Stephanie's clinicians and teachers claimed expertise through their graduate degrees, Marisa did not have a college education. She could not relate to them on that basis, and they could not connect to her experience raising a minimally speaking autistic child. Their class-based condescension made Marisa feel inferior. While the *Most Amazing Story* video did not alter the broader imbalance of power and authority between Marisa and the school district, it did provide an alternative space for Marisa to feel dignified.

David also drew on media texts to highlight what he perceived as a gap between his understanding of Beatriz's AAC needs and those of her teachers and therapists. While these professionals had recommended that Beatriz use Proloquo2Go on the iPad, he felt that the system was too complicated for her to navigate. He poetically described the haphazard layout of the digital app in analog terms. "It is like when you throw word cards on the floor," he explained, gesturing to the carpet, "and what you want, it is right there, but you need to find the word. And I know she can find it, but it won't be that easy because of the way they are." He felt that Proloquo2Go was making it even more difficult for Beatriz to communicate her needs. Said David, "It requires the kids to have some ability in English, like *forming sentences*. Her condition—mental retardation—even though it's not severe, but she has difficulty trying to form sentences."

He compared and contrasted Beatriz's AAC system to those of two media figures: the aforementioned Hawking, and the cinematic portrayal of journalist Jean-Dominique Bauby in the film *The Diving Bell and the Butterfly*.[46] As a young, intellectually disabled, first-generation Latina from a working-class family, Beatriz has little in common with Hawking besides the fact that both use an AAC device to produce speech. David drew on his understanding of Hawking's intellectual pedigree to describe someone for whom using a system like Proloquo2Go would be relatively easy. "I seen the guy, this scientist guy, I don't remember his name. He's very, very intelligent. The one who's studying the stars," said David. "Stephen Hawking?" I asked. "Uh huh!" he continued,

> You know the college guys; they are programming the machine for him to communicate just by looking at it! Just by looking at the screen, whatever his pupil is looking at is a word, and then the machine is repeating the word for him. ... For him it is easier because he is scientist. He can form phrases. He can solve an equation. He can do that because he had the knowledge before. But for Beatriz, it's gonna be hard.

David pointed out that Hawking not only had a masterful grasp of the English language before needing to use an AAC system but also a preexisting network of social and technical support.[47] David felt as though the professionals working with Beatriz had not adequately scaffolded the app for her; she might as well have been using Hawking's technologically and linguistically sophisticated device to communicate instead.

As a point of comparison, David referred to a much simpler, low-tech AAC system that he had seen once in a film that on description, sounded much like *The Diving Bell and the Butterfly*, although David could not confirm the title.

> I watch a movie about this guy who had an accident and he wasn't moving at all, but I think he was a writer. I don't remember her—or was it his?—boyfriend helped him to write the book. And he say, "I'll show you the alphabet. Every time

> a letter of your liking shows up, blink twice." And then after three letters, the girl or butterfly, she knows, she can guess the word. And then that's how she helped the guy to write the book. See? They communicate just by blinking the eyes. So if they can do it by that, why not something else?

David used the language of media to capture what is known among speech-language pathologists as partner-assisted scanning. A better AAC system for Beatriz, David thought, was somewhere between this and Hawking's device. He was not able to use technical, medical, or clinical terms to voice his frustrations, but representations of AAC and AAC users in the mass media provided him with material to do so.

As with Marisa, there are limits to the extent to which David can draw on media in order to improve his child's speech services. In our discussions, David frequently qualified his thoughts on Beatriz's AAC system, with these statements likely shaped in some way by David's perceptions of my own qualifications, education, or racial and ethnic background. "I'm not a specialist in this type of devices or programs or software. I'm not a specialist," he repeated. "This is just my point of view, according to my needs and according to Beatriz's needs." He understood that the dominant culture did not lend much credence to the tacit knowledge he gained from his lived experiences, compared to the codified knowledge one obtains through multiple degrees. To David, Beatriz's school did not understand that her device was too complicated for her to effectively communicate, but he did not feel that it was his place to challenge the school's authority either. Mass media and popular culture became a pathway, albeit a limited one, for less privileged parents like David, Nelson, and Marisa to discuss their children's relationship to the iPad and Proloquo2Go without using specialized terminology.

Conclusion

As with Daisy, whose remarks about Jobs I discussed at the beginning of this chapter, I had to reflect on the meanings that Karun associated with him, her sense of the disability media world, her social status, and my own. When Karun, Pargev, and their family fled the war in Syria for the United States, they stayed temporarily with relatives in Northern California. "We decided, just we need to come back," said Karun. "That's the best place to come." Pargev and his younger brother immediately entered the well-regarded local public school district, and it was there that a teacher told Karun about Proloquo2Go on the iPad. "She said, 'Eighty percent of my children are using it and tremendous language came out.' I couldn't believe," Karun remarked.

What I personally found even more unbelievable, though, was that of all the places to find refuge in the United States, Karun and her family somehow landed in Cupertino, California—home to Apple's corporate headquarters and a school district replete with the latest Apple technology. When I pointed out this incredible luck to Karun, she shared a story further illustrating her feelings of personal connection to the company: "We were walking on Sunday morning with my cousin. She said, 'Do you want me to show you Steve Jobs's house?' I said, 'Yes, I want to see that.' We took pictures, in front of the house. You know that his mom was Armenian?"[48] Thousands of miles from home, Karun found herself linked to a diasporic community of Armenians, of which she considered Jobs to be a member. On the one hand, even in his death, Jobs remains larger than life and a metonym for all things Apple. Yet Karun developed an intimate and localized understanding of the media celebrity. Her conception of the disability media world was tied to her transnational consumption practices. Her appropriation of the figure of Jobs was linked to opportunities for rebuilding her family's life in the United States as well as the maintenance of cultural pride after, as Christian Armenians, Karun and her family faced persecution in Syria.

While Karun can locate a part of herself in Jobs's story, her experiences and those of other less privileged parents receive little attention in the popular press. These media discourses about disability, children, and technology overwhelmingly feature more privileged families, which have greater money, time, stability, and resources. This can potentially mislead less privileged parents who might not have the education and critical media literacy skills to judge the realism of these constructed narratives. In addition, media representations of AAC users generally lack diversity by any measure—including race, ethnicity, gender, sexuality, and class. Given that popular culture is an important way for parents from less privileged positions to gain access to information about AAC users and navigate professionalized spaces, there needs to be a broader range of cultural figures for them to draw on.

Overall, I found three types of parents in terms of their information-seeking behaviors: those who were always looking for information, those who were always surrounded by it, and those not hungering for it exclusively on the Internet. Online spaces geared toward the parents of youth with disabilities may not meet the needs of those marginalized due to their class, cultural, or linguistic backgrounds. Clare Blackburn and Janet Read write, "Information therefore also needs to be available in printed and other forms to ensure that carers who cannot or do not wish to use the Internet are not excluded."[49] Seeing as all the parents I spoke with had some form of Internet access, the issue at hand is more than technical in nature. Claiming knowledge and authority about one's child, their disability, and their technology use requires a degree of social and cultural capital.

The gendered nature of disability media worlds is also important to note considering interactions between class and culture. All parents, although especially less privileged mothers without a college degree, might benefit from free workshops with child care, specifically hands-on tutorial sessions in which they can take physical notes rather than independently learning about technology from online resources in their theoretical "spare time." When engagement with networked media worlds becomes part of the model of ideal maternal involvement—a conception largely shaped by privileged groups—upper- and middle-class mothers are advantaged in wielding fluency of this social language and flexibility with their schedules.

My fieldwork also suggests the need for noncommercial, device-based online groups for parents of all levels of technical expertise. Such groups could support parents whose children might all be using the same type of assistive technology but are dispersed geographically and/or have different diagnoses, and are not censored by corporate interests. Anne, for example, spoke of how she didn't know any other parents whose children were using Proloquo2Go, as her son (who was not Deaf) had been placed in a classroom of mainly Deaf children due to his early use of American Sign Language. Seeing as I had been interviewing lots of other parents whose children were using the same technology as her child, I assured her that there were people nearby with this experience, but wondered how they would ever connect with one another.

In sum, there were clear class distinctions between parents in their participation in the world of disability media. Seeing as most research on the media practices of parents of disabled youth centers on white, upper-middle-class parents, discrepancies in how central the Internet figures into parents' information-seeking behaviors suggests an exception fallacy when only studying those who self-select into such mediated spaces.[50] Peter's comment that "You've gotta be plugged in" works two ways: You've gotta be plugged in to gain access to online information that might help your child, and you've gotta be plugged in to be located by researchers looking for parents primarily through blogs, e-mail lists, social media, and social networking sites.

7 Conclusion

> I don't know what Rachel [the speech-language pathologist] thinks about, or the creators of Proloquo think about, but there should be some kind of study, to go into the kids' homes and see how they are using it. Probably they already did it, but I don't know if they did it with other kids, because there are kids who are *different*. Those who need that program—every kid is *different*.
>
> —David, Beatriz's father (emphasis added)

> I'd be interested to see your research after. I would think that the families you talk to, what we're seeing is pretty much the *same*. It's just different experiences, but we have the *same* thinking about how we want our kid to be. ... You probably have a deadline, but even after you've finished, you'll probably have a part two, because there's a lot.
>
> —Daisy, Thomas's mother (emphasis added)

Over the course of my observations and interviews, a number of parents with whom I spoke, such as David and Daisy, asked me questions about what I was finding out along the way. This metadiscourse about my research was itself revealing. David had a vague sense that his daughter had not been accounted for in professional evaluations of how children use the iPad and Proloquo2Go for AAC. Daisy hypothesized that on a basic level, most parents were like her, and most child users of the iPad and Proloquo2Go were like her son. Both made generalizations (either about every kid or families overall), but the word "different" reappears in David's quote while Daisy repeatedly mentions things being the "same."

On the surface, David and Daisy do not seem so dissimilar. Their lives *converge* as Latino/a parents of nonspeaking children with developmental disabilities in the Los Angeles area. Beatriz and Thomas were both, as Raul's mom, Nina, put it, "born at the best time" (roughly between 2000 and 2010) to be able to take advantage of the iPad as an AAC device at a young age. Each was glad that their children had the opportunity to better communicate their thoughts and feelings. But their lives also *diverge* in profound ways. David is a working-class immigrant from Mexico employed in food service, and Daisy is an upper-middle-class office professional. They are differently exposed to technology in their everyday lives and regularly encounter people with varying degrees of technological expertise in their social milieu.

Their remarks about my research illustrate the tensions that I began describing in the introduction to this book: between understanding parents of youth with disabilities as being united "in a class by themselves" and divided by social class. Throughout, I have shown that these distinctions are not an either-or proposition. The label of "special needs parents" is far from a stable and cohesive category but rather better characterized as an imprecise and always-shifting field, to borrow the language of sociologist Pierre Bourdieu. It is a cultural space governed by internal logics and discourses, with social actors struggling over societal resources. Parents try to maintain or improve their and their child's position in life by accessing different types and varying amounts of capital. The label does not reflect all the ways that parents' experiences differ in terms of how they perceive the diffusion and appropriation of their child's media and technology practices across home and school as well as how they relate to the institutions that directly and indirectly influence the adoption, use, and non-use of those tools and artifacts.

All the children in the families I spent time with had at least basic access to the same combination of hardware (iPad), software (Proloquo2Go), and wireless Internet at home. Yet the technology in and of itself did not "give voice," as mobile devices and networked media are often imagined as bestowing. This book introduces new perspectives on aspects of the iPad, Apple products, and mobile communication devices, including some totally neglected components of the technology, such as mobile accessories like the iPad's physical case. It also touches on dimensions of the hardware and software that have received more focused and sustained attention—such as security passcodes and intelligent personal assistants—but have not been studied in situ or from a sociocultural angle.

At the close, I do not so much offer an alternative to the phrase "giving voice" as I do call on communication researchers to *keep voices attached to people*. Media and society scholars must avoid using disability metaphors to describe lack of voice, especially without engaging disabled people in such conversations. This echoes technology historian Katherine Ott's call for theorists to stop abstracting the idea of prosthetics from the lived realities of prosthetic users. The speaking world stands more to benefit from metaphors about voicelessness and technology than do people with communication disabilities unless the frustrations they and their families voice are heard and acted on.

In utopian visions of technology as a force for leveling the playing field, disabled youth are rarely imagined as active players in the game. The field is difficult to level because it turns on multiple axes of power, with each axis tilting in favor of some and away from others.[1] When used for fun, communication, education, and entertainment, iPads enter into complex systems, conditions, and contexts that reproduce social inequality as well as reward parents who can most effectively mobilize economic, social, and cultural capital in its embodied, objectified, and institutionalized forms to support their children. The privileges that nondisabled upper- and middle-class children gain by being on the right side of the "home technology divide" extends to upper- and middle-class youth with disabilities along with the additional resources they have at home to support their use of assistive technologies.

In this conclusion, I provide an overview of the various ways in which more and less privileged families of youth with communication disabilities both are and are not alike, and summarize the contribution that this book makes to conceptual notions of voice as well as theoretical understandings of capital and structural inequality by way of intersectional analysis, centering on class and disability. I emphasize the importance of examining multiple and overlapping systems of power and oppression that privilege certain parents of youth with disabilities over others. I then take a step back to examine moments in which more privileged parents brought intersectionality into the forefront of our conversation. Their self-reflection is crucial, but it is also limited in its impact without concurrent systemic change. In that vein, I propose practical applications for how this research can be taken up by various audiences: parents, health and education practitioners, technology developers, policy makers, and media content creators. Lastly, I suggest areas for ongoing research in this rapidly developing space of voice, communication, and technology.

Similarities, Differences, and Intersectionality

Mobile technologies, largely thought to universally empower people by "giving voice to the voiceless," particularly individuals with disabilities, are still subject to disabling structural inequalities we often do not hear about, and as a result, are assumed not to exist. I did not enter into this project looking for the iPad, but after conducting initial fieldwork on both dedicated and nondedicated AAC device use in family homes, encountering increased attention to Apple in the US press, and overhearing anxious discussions about the iPad among assistive technology professionals, I found the object to be a key site of cultural change through a grounded approach to understanding technology, society, and disability.

This book is thus framed by five main questions about the role of the iPad as a mobile communication technology: What is voice? (chapter 2); What is a mobile communication device? (chapter 3); What is an iPad for? (chapter 4); What does it mean to communicate with an iPad? (chapter 5); and, How do media shape understandings of the iPad? (chapter 6). Although I expected perpetual difference between families across class in how they answered these questions, I did not always find it. While the "null hypothesis" is underreported in quantitative social science, I employed qualitative fieldwork, and through it, discovered that more and less privileged parents were not radically different from one another, but that different differences mattered when understood in relation to broader social, cultural, economic, political, and technical contexts.

In terms of similarities, I found that more and less privileged families of youth with communication disabilities shared a number of key experiences and perspectives. In chapter 2, I discussed how parents had a more nuanced understanding of their child's voice than dominant discourse in the news media surrounding voice and AAC suggests. I illustrated in chapter 3 how the physical case that covers the iPad—an aspect of the technology that might otherwise be considered disposable and trivial—became an important point of how parents saw their child, their child's device, and the social structures in which their child's technological practices were embedded. Chapter 4 focused on the majority of families in the study that wound up with more than one iPad in their home—a phenomenon that occurred across income levels. Chapter 5 showed how parents of youth with disabilities can feel as though their child is discounted by society, but that their child's use of media and technology provides a new pathway for reclaiming dignity and recognition of their abilities. And in chapter 6, I revealed how all parents participate in disability media worlds in some manner. Across class, parents found opportunities for social and emotional support from online communities, valued the stories of nonspeaking children that they discovered through mainstream and user-generated digital media, and also felt that the most famous AAC users in mass media were not necessarily the most appropriate role models for their child.

But there were major differences among parents too. Chapter 2 brought to light the ways in which voice output communication aids and synthetic speech technologies are gendered and raced through sociotechnical practices. In chapter 3, I detailed how less privileged parents had little choice but to be tied to the whims of school districts, which seemed to care more about protecting the technology they are legally mandated to provide than serving the child whom those provisions are supposed to help. Chapter 4 revealed how schools and therapy providers embed middle-class values into the language they use to talk about the iPad, and that this inherently benefits more privileged children and parents in educational and clinical settings. In chapter 5, I showed the limits of popular culture and new media as a way for working-class nonspeaking children to seek alternate modes of expression in view of the other ways in which more powerful entities undercut their competencies. And I demonstrated in chapter 6 how presumptions of uniform access to information and proficiency in technology among families can marginalize less privileged parents.

One way to understand these varied experiences is through Bourdieu's conception of capital.[2] Researchers of parental participation in children's schooling, special education, technological practices, and home media use have all drawn on Bourdieuian field theory.[3] While I started from his notions about distinction, they were not my end point. One limitation of Bourdieu, write Patricia Collins and Valerie Chepp, is that "his work focuses primarily on a system of class domination and therefore is limited by its lack of an intersectional analysis."[4] This book draws field theory deeper into interrogating the boundaries, relationships, locations, and dislocations of disability and technology in everyday life. Beyond a single population, hardware, or software, this book has highlighted the importance of examining intersecting differences in how individuals and families make sense of technological and cultural shifts relative to overlapping systems of privilege. I have linked otherwise largely abstract notions of voice with specific passages about privilege and agency throughout the book, revealing particular kinds of indignities and controls imposed on working-class youth with disabilities and their parents by interconnected bureaucratic systems.

Recognizing Privilege

Hawking made headlines in May 2015 for calling attention to how structures of power work to isolate less privileged disabled youth from opportunities for upward mobility. At an event to mark his fiftieth year as a fellow of the University of Cambridge's Gonville and Caius College, Hawking reflected on escalating cuts to higher education spending in Britain and their impact on educational opportunities for students with disabilities. He remarked, "I wonder whether a young ambitious academic, with my kind of severe condition now, would find the same generosity and support. ... Even with the best goodwill, would the money still be there? I fear not."[5] That support not only included scholarship money during Hawking's doctorate but also funding for significant infrastructural adaptations to his student apartment building, including the installation of an elevator.

While Hawking did not speak further on the issue, his comments appear driven by a fear that the proposed UK spending cuts would also serve as a disincentive for talented yet less resourced disabled youth to complete their degrees or even apply in the first place. The cuts would be less of a deterrent for more privileged students, who might be able to mobilize additional resources to study and carry on their research. Hawking appears mindful of the advantages that he has been granted through public and private means over the course of his career, and is using his powerful position as perhaps the most famous scientist in the world to lend his voice and advocate on behalf of those disabled youth without the same guaranteed level of support.

I would be mischaracterizing more privileged parents to say that they too were completely unaware of their and their children's advantages. In fact, there were moments in my research in which these parents made an explicit point of recognizing the material and symbolic resources that they, like Hawking, had at their disposal. For example, toward the end of my interview with Luke's parents, Rob and Debra, they asked me how I had arrived at my research topic. I mentioned that while I had difficulty pinpointing one specific reason, the fact that my sisters and I have all ended up working professionally in the space of disability (one as a speech-language pathologist, and the other as an occupational therapist) suggested some common motivations.

I mentioned how, due to this shared background, I was able to help my sister write her application to graduate school for communication disorders and sciences—an essay that focused on how the ability to communicate was inherently a social justice issue. "That's surprising that someone would characterize it as a 'social justice issue,'" replied Rob, to which I responded, "Hmm?" I sensed that I had transgressed some sort of definitional boundary. "It's just the term 'social justice' is a kind of [a] dramatic term. It conjures up images of people being held down," Rob explained. He associated the term with stigma and oppression as well as the experiences of historically marginalized ethnic groups: "You can argue being an autistic in a Spanish community or something is a social justice issue. I can get that. They routinely get screwed on services. But I don't think Luke is getting gypped [*sic*] on services relative to his speech issue because he doesn't speak."

Rob rebuffed the suggestion that his son was "being held down" by society in any manner. In his view, Luke's status as a white, upper-middle-class boy largely shielded him from being disenfranchised. Rob emphasized that his son was different from autistic youth in "a Spanish community," a phrase he associated with economic disparity, because Luke was not lacking in access to social services. To Rob, the iPad with Proloquo2Go was not "alleviating an *in*justice" but instead "broadening the options, giving him a better shot at [communication], which is great. One of my hot buttons I guess!" he said with a nervous laugh.

Another upper-middle-class white father, Peter, discussed how he was in a different position than low-income Latino/a/x parents of disabled youth. Peter, Alice, and their son Danny live in Ventura County, at the southern end of California's agriculture-rich Central Coast. California has by far the largest number of migrant farmworkers of any state, and Ventura is one of the major counties in which these individuals live and work.[6] Peter had been complaining about his difficulty in accessing pricey, proprietary, written materials produced by AAC app companies when he took a moment to reflect on his relative privilege:

> In Danny's classroom, probably half the parents in there are professionals on one level or another, and the other half work in fields. No offense against the field workers, but they don't have the education or even just the English to digest this, but they've got the same kid I've got. Why are they not able to get a version of this for free that helps them? Just because they don't speak English doesn't mean they don't love their kid. They want the best for their kid too.

For Peter to have the time to read and use support documents and manuals, with their specialized technical language in English, to best support Danny in using the iPad as an AAC device not only required significant economic capital but cultural capital as well. While "they've got the same kid I've got," according to Peter, education and language sets the classroom parents apart from one another.

Parents like Rob and Peter in no way speak for less privileged families of nonspeaking youth in absentia, just as Hawking does not speak for all AAC users. They do, though, highlight the complex ways in which access to communication technologies both is and is not a social justice issue for those in the disability community who are "partially privileged" due to their race, gender, age, sexual orientation, or social class.[7] More privileged parents of nonspeaking youth can at times have more in common with similarly class-privileged parents than with less privileged parents of nonspeaking youth, and vice versa. It also matters whether or not parents think of AAC as a social justice issue because many adult AAC users do think of it as such. Yet mass media give parents a platform to speak about disability issues more often than they do disabled adults.

Coming to a personal framing of communication access as a social justice issue depends, then, on a complex mix of temporal, historical, and political factors as well as parents' subjective interpretation.[8] As noted in chapter 1, for instance, the class status of Karun and her husband significantly shifted on their move from Syria to the United States, shaping the social, economic, and cultural resources available to them to care for their autistic son, Pargev, and support his use of new media for play, learning, and communication. Oppression and privilege are not simply additive or cumulative but rather must be read in relation to a set of shifting, contextual variables and multiple identities.[9]

Implications and Recommendations

Over the course of my fieldwork, I began to notice the great opportunity and responsibility I had to translate my findings to different stakeholder groups. I had initially assumed, quite wrongly, that the insights I was gaining were already evident to teachers and therapists. For instance, in conversation with a school district AAC specialist and her assistant, I asked them if they found that there was a direct relationship between the AAC device going home and a child using it more in school. The specialist noted that it depended on, "Did it ever come out of the backpack, when it was home? I don't know, because we're not at home. We can't see." The assistant also spoke about this guesswork among school-based practitioners with respect to home life. "When the device goes home," she said, "we feel that maybe there is some interaction with the device, with the family. I guess in the initial stage maybe. We're hoping, but we don't have the ..."

That ellipses hung in the air like a giant fill-in-the-blank. They didn't have the time to make additional visits over months or years to family homes. They didn't have the school funding to pay them overtime for such work. And they didn't have the information I was seeking. Based on the insights I gained through my unique position and qualitative methodology, there are a number of recommendations and implications of this work for practitioners as well as parents, technology developers, policy makers, and the mass media.

Practitioners

I encourage clinicians, therapists, and teachers to develop greater reflexivity regarding their own class biases in how they approach the iPad and disabled children's use of it. It is impossible to separate professional beliefs about the "proper" use of an iPad as a piece of assistive technology from their own personal or even lifelong beliefs about the "proper" role of new media in children's lives overall. Upper- and middle-class practitioners who cannot see the viewpoints of less privileged families risk further isolating them. I want to caution, though, against solely blaming schoolteachers, therapists, or administrators either, for they are making do with high-stakes demands on their own time and significant budgetary constraints.

These class biases exist not just among individuals but also at the institutional level, where there is much assumed about how parents seek information regarding how best to promote their children's technology use. While high-earning parents with a college degree may welcome online tutorials as a way to supplement their own technology training, this also isolates many caring parents for whom "just going to a website" is a significant challenge. It is not that these parents do not care—but as both caregivers to learners and learners themselves, they bring their own educational experiences to the table.

New approaches to parent technology training might take a cue from disability studies, specifically the pedagogical principles of Universal Design for Learning.[10] Grounded in neuroscience research, Universal Design for Learning is a framework for optimizing instructional goals by building variability into learning environments. For our purposes here, this could include presenting parents with information and content about the iPad and Proloquo2Go in a variety of modalities, and using different engagement techniques to stimulate parents' interest and motivation. The content of this parent technology training should also be expanded to address gaps that this study revealed. Parents were unevenly aware of the range of synthetic voices that AAC devices offer, options for backing up the software wirelessly through cloud computing, and how data in the app are stored and managed by school districts.

Most important, practitioners need to interrogate the direct and indirect ways in which they can unfairly make parents and youth with disabilities prove their competence with technology in order to benefit from it, and the ways in which they can undercut competency as human beings in the process. This lack of presumed competence especially burdens less privileged families because it takes significant social, cultural, and economic resources to provide proof of it in a form that appeals to practitioners while meeting their own professional and legal obligations of record keeping and data logging.

One way to remedy this is for practitioners to understand the iPad more holistically: as a cultural object besides a medical tool. This includes taking notice of iPad cases not as a cosmetic afterthought but instead as something that can impact whether or not the technology—over which so much time and money has been spent—actually gets used by children and benefits them (chapter 3). It also involves clinicians and therapists understanding media and technology in children's lives more broadly—and how children draw on this material to communicate (chapter 5).

On a macro level, these biases could be addressed through more time allotted in the graduate curriculum for medical professionals who work with child AAC users (e.g., speech-language pathologists, occupational therapists, and physical therapists) to learn more broadly about children's media and technology use. A sentence or two about how families differ in terms of their views on technology, and the role this plays in clinical interactions, might be incorporated into the ethos of professional organizations such as the guidelines of the National Joint Committee for the Communication Needs of Persons with Severe Disabilities, and the online resources that the American Speech-Language-Hearing Association provides for speech-language pathologists on the topic of working with culturally and linguistically diverse individuals.[11]

A way to address class bias on a more micro scale might be an intake survey or informal conversation between practitioners and parents early on during their time working together so that each can become better acquainted with one another's expectations—not only of the iPad and its use as an AAC device, but their preexisting values, beliefs, and habits about the role of new media in children's and families' lives. Researchers in both communication science and communication studies could collaborate on constructing such an assessment form.

Parents

Stephanie's dad, Nelson, spoke proudly of how his wife, Marisa, had "been very persistent and very adventurous as to, 'I'm just going to do this on my own. I'm going to go to the library, get books, and find resources, and do whatever it takes to help our daughter.'" The notion of a DIY ethos of US individualism generally evokes a sense of agency, particularly with respect to cultures of making, hacking, and tinkering with technology.[12] But when phrased by Nelson as "do whatever it takes" and "do this on my own," this independence more accurately reflects abandonment due to neoliberal logics of care and a limited range of resources available to working-class parents. More privileged parents of youth with disabilities that engage in "DIY parenting" practices also receive invisible support from multiple sources, such as the diffusion of information through extensive and reliable social networks.

At a unique time in US history in which caregivers are increasingly isolated from community and family support, the best thing that parents can do is to find alternative ways to "do-it-together." That alternative model of community-supported child-rearing will look different for a given family, but "together" in this sense should be rooted in the sociopolitical space of disability. It involves the mentorship and wisdom of adult users of AAC devices along with disability advocacy organizations with a vested interest in AAC, such as the Autistic Self Advocacy Network and the International Society for Augmentative and Alternative Communication. This coalition building can provide adult AAC users, young AAC users, and their parents with a powerful sense of social and cultural belonging.[13] Parents' accounts do not stand in for the self-advocacy of adults with disabilities, who refuse to have their voices drowned out in a continued struggle for recognition.

None of this is to say that more privileged parents of youth with disabilities should refrain from drawing on their resources in order to free up more opportunities for less privileged parents. Notes Colin Ong-Dean, "Even when they are most narrowly focused on the well-being of their own families, privileged parents are, after all, trying to help not themselves, but their children, and they are dealing with laws and institutional practices that limit what can be done."[14] In turn, though, policy makers, practitioners, technologists, and media producers must write legislation, provide services, build tools, and craft content in a more egalitarian as well as culturally sensitive way to meet the needs of more families.

Technology Developers

This book also illustrates the need for those who design communication technologies for use by those of all backgrounds to be attentive to disability, and for those designing for disability to be attentive to other forms of difference. There is no straightforward resolution to designing technologies for children and families across both social class and different disabilities.[15]

While class is a persistent theme in communication, sociology, and anthropology, this has been less true in research on human–computer interaction and computer-supported collaborative work in a number of areas, detailed below.

Voice and Diversity One specific area of computer engineering with significant room for improvement is in the design of synthetic speech technologies, and how assumptions about race, culture, and gender are built into systems for technologized voice. It should be noted, however, that synthetic speech systems are far from the only technologies for expression and communication that inscribe race and gender, silently and invisibly. Cinematographic tools, for instance, have historically been calibrated to film white skin, using models of light-skinned women in colorful dresses (also known as "China Girls") to achieve color balance. Compensatory practices are similarly required in order to reproduce nonwhite skin colors on screen in film and television.[16]

Besides our words, our voices say so much about us. The iPad with Proloquo2Go structures the possibilities of speech and identity. Nonwhite and female voices are largely silenced through synthetic speech systems, and have been since they originated in the largely male- and white-dominated spaces of military and academic research in the twentieth century. More personalized AAC systems enhanced by machine learning, artificial intelligence, and contextual awareness offer new options, although these are not yet widespread.[17] Expanding the range of human voices embedded in machines not only has implications for assistive speech technologies, and those who communicate through them daily, but all ubiquitous and artificially intelligent systems through which machines and humans converse too.[18]

The AAC community is taking the lead in driving such innovations. One such initiative is VocaliD, a start-up launched by Professor Rupal Patel, a colleague at Northeastern University. VocaliD started a crowdsourcing project to create the world's largest collection of standardized speech recordings from English speakers of all ages and backgrounds, with the goal of crafting more personalized synthetic voices. The goal of diversity runs up against other structural limitations, though. In order to become a speech donor, one must have three to four hours of spare time to record their speech, access to a steady and strong Internet connection, a computer or smartphone with a microphone and the Chrome browser, and a quiet location in which to record. To reach the widest population possible, novel outreach methods and forms of compensation should be considered for those less privileged potential voice donors.

Data and Privacy High-tech AAC devices keep track of everything a person says, but in an untraditional way: by storing digital records of symbols or text converted into speech output. Individuals who use high-tech AAC devices thus have unique perspectives to share on issues that concern Internet and society scholars, such as data surveillance, countersurveillance, and personal privacy. They can also make valuable contributions to important discussions about the complicated ethical and technical entwinement of listening and speaking machines that purport to assist humans in daily living.[19]

Besides hearing the technologized voice or hearing the AAC users, does one hear a person who can make decisions for themselves? Communications and disability consultant Michael B. Williams writes,

> Augmented communicators have a cultural history of being abandoned without communication in human warehouses and "special" schools. They may have had personal experience offending someone whom they rely on for care or services. "No. Don't touch that. This is mine. Leave me alone," are not foremost in the minds of many augmented communicators. Not many augmented communicators are going to risk losing the assistance and support they have been struggling for in order to gain some privacy.[20]

Only in the past few decades have disabled individuals won societal recognition as being able to make decisions about their own assistive technologies and gained legal influence in shaping the policies that regulate these devices. Having a voice, and the role of technology in exploiting that voice, must be understood in relation to other forms of exploitation that characterize the systemic struggles of historically marginalized individuals, including those with disabilities. They are not passively given voices by able-bodied individuals; disabled individuals are actively taking and making them despite structural inequality.

Talking and Listening Machines There exists a good deal of valid anxiety about the privacy and security issues raised by the Internet of Things (i.e., wearable technologies, connected household appliances), and the data collected by conversational Internet-connected objects with synthesized speech output such as Mattel's Hello Barbie and iPhones with the voice-activated personal assistant Siri. These debates should also be contextualized within the larger history of AAC systems, as these technologies innately involve the practice of speaking *through* and being in conversation *with* objects.[21]

Artificially intelligent talking toys and their intersections with AAC might also influence the design of educational technologies more broadly. While these toys do not necessarily promote language learning in all children, and can in fact be detrimental for typically developing children, talking toys might "become stepping-stones to conventional text-to-speech devices" for children with complex communication needs.[22] Exploring AAC through the lens of electronic toys might allow for greater insights into language development in the digital age.

Tone of Synthetic Speech Compared to talking toys such as Hello Barbie, which speak through digitized speech recorded by voice actors, most voices in AAC systems have a neutral reading style. Yet this imagined neutrality does not reflect the full span of human expression through voice. Children and adults do not generally speak as if they reading dispassionately. Tone of voice is not secondary or optional but rather essential for being understood by others. There is the need, then, for more creative, human-centered design research into how tone of voice in AAC devices might be expanded beyond, for instance, the ExpressivePower option in Proloquo2Go in a way that best fits the needs of individuals across the life span, and in a manner that reflects the contemporary communication environment in which individuals coconstruct meanings through mobile media with multiple conversation partners.

An initial entry point for such progress, proposed by Graham Pullin and Shannon Hennig, is a "social model of tone of voice." By this they emphasize the relational and social aspects of mediated voice in addition to a specific individual's emotionality. There is not just one way to say "yes" to another person but instead at least seventeen by their count. Pullin and Henning conceptualize tone in synthetic speech systems as consisting of four dimensions: vocal qualities, emotional state, social context, and conversational intent.[23] This social model of tone of voice has implications not just for assistive speech aids but rather all mediated communication systems used by people with and without disabilities that augment or serve as an alternative to embodied oral speech.

A social model of tone of voice might inform the design of synthetic speech systems that translate the alternative textual descriptions, or "alt text," that accompany emoji symbols from symbol-to-text into text-to-speech. Emoji are wildly popular worldwide and relay emotion exclusively through visual communication.[24] But what should emoji—with alt text titles such as "Heavy Black Heart" or "Pile of Poo," as denoted by the international software standardization group Unicode Consortium—actually *sound* like in different languages when spoken through synthetic speech systems? When Apple VoiceOver and other screen readers speak this alt text, it is aurally conveyed in the "neutral reading style," even though this style does not necessarily match the meaning that the user intended to convey by sending the emoji image in a text message.

In written text, emotion is formally communicated through punctuation, as in question marks and exclamation marks. There is no "sarcastic mark" or "sincere mark," however, that accompanies the alt text for emoji. The complex task of translating symbol-to-text and text-to-speech not only applies to emoji but emoticons as well. Text-to-speech systems cannot properly convey the collection of punctuation signs that comprise the "shruggie" emoticon—¯_(ツ)_/¯—as vocal intonation. Considering the fact that people with communication disabilities have access to more types of communication technology than ever before, and that some individuals on the autism spectrum employ emoticons as "social captioning" for understanding social cues in mediated conversation,[25] we need culturally richer symbol- and text-to-speech systems that tonally represent the vocal complexity of today's mobile and mediated world.

Online Communities This book also points to the need for more kinds of digital spaces for parents and other caregivers of AAC users to gather and share resources. This is in addition to online support groups based on geographic location or specific disability. While the majority of parents who participated in this project had children on the autism spectrum, it is important not to lose sight of the fact that children may be unable to speak, or have significant difficulty doing so, due to other developmental disabilities and rare disorders. When the iPad and AAC become synonymous in the news media with autism, the specific needs of nonspeaking children with other disabilities and their families risk being overlooked in other media and medical spaces.[26]

Policy Makers

The announcement of Hillary Clinton's wide-ranging autism initiative, proposed in January 2016 during her campaign for the US Democratic Party's presidential nomination, suggests that a subset of US policy makers are starting to take a closer look at the everyday issues that directly impact individuals with communication disabilities, including access to assistive technologies. As part of a broader enforcement agenda on protecting the rights of autistic individuals, Clinton identifies "variations in state law and enforcement" as a barrier to AAC use, especially when these tools "are made available as apps in consumer mobile or tablet computing technology," or when students lose "access to effective tools when they move between schools."[27]

Clinton's proposed solution is "stronger oversight and enforcement," including "an evaluation of access to assistive technology in the state autism scorecard."[28] While a significant number of AAC users are also autistic, I caution legislators not to conflate AAC and autism in how resources are provided to AAC users, lest those with other disabilities become underserved as a result. Beyond the capability to use these communication tools, accessible technology experiences require other kinds of social, cultural, economic, and political infrastructures.[29]

For instance, health insurance and local school district assistive technology policy both need to be revised to better reflect the new landscape of AAC devices. The total lack of third-party insurance coverage for tablet-based AAC devices inherently privileges upper-income parents who can afford to pay for them or pay property taxes in well-resourced school districts. This creates different tiers between families. There were many examples of less privileged parents being locked out from using other potentially useful and educational apps on the school-owned iPad, or frequently locked out from using the device at home at all if they accidentally violated school protocol, such as charging the device using their home computer.

While I understand that schools are supplying assistive technologies first and foremost to support children in being able to access the curriculum, those young people who use AAC devices do not only learn or communicate at school. Current policies neglect the home as a place of enrichment, and that children might benefit from using tablet computers as AAC devices as well as for other communicative purposes such as e-mail or social media. Instead of indiscriminate policies, the dedicated or nondedicated use of tablet computers as assistive technologies should be treated on a student-by-student basis in order to maximize these children's individual opportunities for communication, interaction, and inclusion in society.

Increased funding is also needed for those in the public sector who support AAC users in the community. Already-underpaid public school special education teachers and speech-language pathologists in the United States must generally pay out of their own pockets to receive training on advancements in AAC devices. Furthermore, key partners in local neighborhoods, such as police and fire departments, require additional training in how to communicate with people who do not express themselves primarily through oral speech and may be in critical need of their services.

Mass Media

Lastly, there are vast improvements to be made in the reporting and representation of disability and technology issues in the media, and especially around AAC and AAC users.[30] The book's discussion of using children's engagement with new media and popular culture to understand their experience of disability is a novel insight. While this topic has received some popular attention (namely through journalist Ron Suskind's memoir, *Life, Animated*), this book is the first to offer a systematic study beyond a single family's experience, and situate such work within a broader social, cultural, and political context. It puts family media use in conversation with dynamics that have a strong influence on communication—such as race, generation, education, and gender—especially for users with disabilities. While disabled individuals are disproportionately underrepresented in fictional and nonfictional stories in news and entertainment, it is important to not just focus on quantity.[31] There is a need for representations of more kinds of AAC users with a broader array of backgrounds, not just white, upper-middle-class individuals whose visibility is already heightened by their access to social, cultural, and economic resources.

Future Research

There will, as Daisy noted at the beginning of this chapter, probably be "a part two" of the research presented in this book, if not more. This work on voice, communication, and technology invites several potential directions for future scholarship across multiple disciplines, including but not limited to media and communication studies, information studies, disability studies, science and technology studies, communication science and disorders, learning sciences, educational psychology, and sociology.

Informatics researchers may be particularly interested in the spectrum of information-seeking behaviors developed in chapter 6—always looking for information, always surrounded by information, and not hungry for information. The latter group was not uninterested but instead pushed away the metaphoric plate of information when served up exclusively in ways that were unappealing, inaccessible, or misaligned with their social and cultural needs. This spectrum may also be applicable to other work being undertaken more broadly by communication and information researchers on information seeking across digital and nondigital media.

Those who specifically study the relationship between the infrastructure of mobile media, political economy of technology, and societal inequality might take up the tension that I identified in chapter 4 on what as well as who was being *locked in, locked out,* and *unlocked* based on overreaching technology policies, and trace how these dynamics change related to various kinds of technology use and non-use in different social, cultural, and political contexts.

More research is broadly needed at the intersection of communication studies and communication sciences.[32] It is unclear how mobile technologies are being taken up as AAC devices among nonspeaking young adults and adults with developmental disabilities who are no longer served by school districts for their assistive technology needs as well as by adults with acquired disabilities. Although I study and write largely about children, there may also be similarities among adults across the socioeconomic spectrum in terms of the possibilities and challenges posed by the convergence of assistive and mainstream networked mobile communication technologies. This includes how they identify or disidentify with their synthetic voices (chapter 2), and how they use mobile social media such as Facebook or Instagram on their AAC devices to participate in disability media worlds (chapter 6).

Among children and media researchers, there needs to be further market analysis of how special education apps fit more broadly into the ecology of children's technology as well as how they are situated within the cultural history of children's software.[33] More quantitative and qualitative work is also required in order to understand how youth with various disabilities and their family members incorporate all sorts of media and technology into their lives. Feminist disability theorist Alison Kafer writes that in the United States, "the always already white Child is also always already healthy and nondisabled; disabled children are not part of this privileged imaginary except as the abject other."[34] The media-using child is not only assumed white as default but able-bodied, too, thereby othering youth with disabilities from normative constructions of childhood.[35] This exclusion ultimately obscures the multifaceted nature of disabled children's media use, and how disability and specific disabilities intersect with race, class, ethnicity, nationality, language, and gender.[36]

One specific area for continued research is the everyday digital media practices of autistic youth across both the autism and socioeconomic spectrum.[37] According to the US Department of Health and Human Services' Centers for Disease Control and Prevention, an estimated one in forty-five children are on the autism spectrum.[38] One interpretation of this statistic (and there are many) is that to thoroughly study contemporary US childhood is to also study autism.

Autistic youth, particularly those with Asperger's syndrome, are often characterized as being naturally geeky or tech savvy.[39] Various technology companies such as Microsoft and SAP have launched high-profile employment initiatives to recruit autistic workers for detail-oriented tasks such as data entry and software testing. The *New York Times* describes this as "The Autism Advantage," implying that autism is both an advantage in the workforce and advantages companies in the marketplace.[40] This flips the usual script of disabled people as in need of repair and technology as the key fix; instead, "high-functioning" individuals on the autism spectrum are figured as technology industry solutions.

While workplace discrimination of individuals with disabilities is all too frequent and their underemployment rates are alarmingly high, a singular focus on this particular solution is misguided. Employment and empowerment are not one in the same; many disabled people seek the right not to work as their only pathway to social value.[41] There are other aspects of the social and cultural infrastructure that need extensive repair work, or were never functional to begin with. As with the "digital natives" fallacy, the perception of autistics as technology savants can mask complex intersections with class, gender, and race in the history of autism. Leo Kanner's conception of the condition in the mid-twentieth century as being more common among white children of the upper and middle class was based entirely on referral bias and selective sampling.[42] In the twenty-first century, those autistic students who go to college and find postsecondary employment are usually from middle- and upper-income families with college-educated parents; less privileged autistic high school graduates from lower-income backgrounds and first-generation parents are less likely to attend college at all.[43]

Discourses about a naturalized relationship between autistic youth and technology can also obscure individual variations in children's preferences. Daisy, for instance, said her son, Thomas, "never really touches the computer. Thomas is not one of those. He's not into the PS2 [PlayStation 2]. ... He's not into any gadgets. He's not a 'techy' person. Some kids you give them any game on the computer—he's not like that." While there has been some survey research on autistic children and adults' media and technology use as well as parental attitudes toward their media engagement, there is a significant lack of ethnographic work and qualitative scholarship that accounts for working-class, nonwhite, and/or nonmale autistic children along with what media and technology use means to them, their families, and their futures.[44]

Conclusion

This book establishes disability as a key topic in media, communication, cultural studies, Internet studies, sociology, and science and technology studies, with its framework braiding together Bourdieuian theories of class, distinction, and education. While this book is one of the first to cross over between media/communication and disability scholarship, it is firmly committed to improving the actual lives of people with disabilities by making recommendations for a number of stakeholders: from parents, to practitioners, to media, technology, and policy makers.

The role of mobile technology in shaping individual and collective well-being is a topic that concerns many disciplines. As it pertains to disabled individuals' technology use, discussions about well-being too often remain bounded within the health sector, or secondarily, the field of education. But living with disability is about more than medicine or rehabilitation; it is profoundly rooted in cultural, social, political, historical, and economic factors. Communication technology scholars are trained to speak to these dynamic forces, and have much to contribute to advancing digital and human rights by speaking alongside and in continual conversation with disability communities.

The deployment of the iPad as an assistive device for nonspeaking children is shaped by the interconnected systems within which families of children with disabilities in the United States have been entwined historically. Those living with disability are sometimes included and at other times excluded from the generalizations that communication scholars make about the place of new media technologies in everyday life as well as a force for democracy. AAC devices serve specific purposes, but just like other mobile communication technologies, they are shaped through social practices and have broader societal consequences.

As such, it is crucial that we avoid using technologically and socially determinist language as well as the notion of "giving voice to the voiceless" to explain the use and adoption of tools that generate voice output by individuals with complex communication needs. Speech acts construct our shared notions of reality, but so does the language that we use to talk *about* speech. Such meanings depend not only on those whose turn it is to speak but on their conversation partners as well. This book reflects not only my personal voice as a researcher and writer but also respects the reflections of parents and children with disabilities, who are in many ways not always associated with being experts or authorities on their own lives. Their voices are woven within this book, in dialogue with and counterpoint to my own.

Technologies that give voice to people can also create gaps between them. There are no such things as haves and have-nots when it comes to voice. As with access, voice has many distributed dimensions. Ultimately, not only can communication research serve the interests and rights of disabled citizens, but disability and the stories of nonspeaking individuals can in turn improve understandings of voice in communication research as well as society writ large. They have a lot to say if the rest of us can figure out how to listen.

Methods

This section details my methodology in undertaking qualitative research with parents of youth with complex communication impairments. I discuss how I determined participation criteria, recruitment challenges, study design, mode of analysis, and research limitations.

Participants

Table M.1 provides descriptive data for child and parent participants. There were three criteria for parental participation. First, parents had to have at least one child with the following characteristics: the child was between the ages of three and thirteen at the start of the research, embodied oral speech was not the child's primary mode of communication, the child used the iPad and Proloquo2Go as an AAC system, and the cause of the child's communicative impairment was a developmental disability. Second, at least one parent spoke and understood English at such a level as to not need an interpreter. Third, parents had to live in Los Angeles County, Ventura County, or Orange County, California.

Table M.2 presents the demographics of the child participants. Their disabilities mostly included autism (fifteen children), but also cerebral palsy as well as rare genetic and chromosomal disorders. Throughout this book, I have removed any mention of these specific disorders in order to protect participants' anonymity. It is estimated that 30 percent of individuals diagnosed with autism are not able to reliably produce oral speech.[1] The exact proportions of non- or minimally speaking youth with developmental disabilities is unknown, nor how this categorization might shift over the life span.[2] No medical files or school records were consulted during my research, and parents provided the language used to describe their child's disability.

Table M.1

Descriptive Data of Child and Parent Participants

Child *N*	20
Age (median)	8
Male	16
Female	4
Disability	
Autism	15*
Cerebral palsy	3
Rare genetic or chromosomal disorder	3*
Race	
White	11
Nonwhite	9
Latino/a/x	3
Mixed race	3
Asian	2
Black or African American	1
Household income[a]	
≥ $100,000	8
$99,999–$50,000	2
$49,999–$25,000	4
≤ $25,000	1
Language other than English used at home[b]	
No	10
Yes	7**
Parent marital status	
Married	18
Single	2
Father's education [c]	
College or advanced degree	11
Some college, high school, or less	5
Mother's education [c]	
College or advanced degree	11
Some college, high school, or less	5

* One child was both autistic and had a rare disorder.

** These languages included Spanish, Armenian, and American Sign Language.

[a] Five families did not disclose their household income.

[b] Three families did not disclose their linguistic background.

[c] Four families did not disclose either their father's or mother's education level.

Table M.2

Demographics of Child Participants (Order by Date of First Parental Encounter)

Name	Age*	Gender	Race/ethnicity	Disability	More/less privileged
Paul Michael	8	M	White/Armenian	Autism	Less
Beatriz	10	F	Hispanic, Spanish, or Latino/Mexican, Mexican American, or Chicano	Cerebral palsy, epilepsy	Less
Nash	3	M	White	Cerebral palsy	More
James	8	M	Asian/Korean	Autism	Less
Madeline	9	F	Asian/Filipino	Autism	Less
Thomas	11	M	Mixed race (Asian; white)/Filipino	Autism, intellectual disability	More
Stephanie	10	F	Hispanic, Spanish, or Latino/Mexican, Mexican American, or Chicano	Autism	Less
Pargev	13	M	White/Armenian	Autism	Less
Raul	5	M	Hispanic, Spanish, or Latino/Cuban; Mexican, Mexican American, or Chicano	Autism	More
Talen	6	M	Mixed race (black or African American; Hispanic, Spanish, or Latino; white)/Puerto Rican	Autism	Less
Luke	13	M	White	Autism	More
Danny	6	M	White	Autism, epilepsy	More
Isaac	8	M	White	Autism	More
Kevin	13	M	Mixed race (Asian; white)/Japanese	Autism	Less
Eric	6	M	White	Rare chromosomal disorder	More
Chike	7	M	Black or African American/Kenyan	Autism	More
Cory	4	M	White	Rare genetic disorder	More
Sam	8	M	White	Autism, intellectual disability, genetic disorder	More
River	7	M	White	Cerebral palsy	Less
Moira	10	F	White	Autism	Less

* Child's age at the beginning of the research.

While ages three to thirteen might be a rather large range for traditional studies of children and media, there were a number of reasons why the study was bounded in this manner. First, chronological age is an arbitrary marker when one considers that the developmental age of this group of children tends to be younger in certain domains (e.g., physically or cognitively). While Elmo tends to be beloved by preschoolers, for example, he was one of thirteen-year-old Danny's favorite media characters. At the other end of the age range, three-year-old Nash was a big fan of Miles Davis and other jazz musicians. "Age appropriate" media is not necessarily appropriate for every child.[3]

Second, the age range of three to thirteen was also based on the significance of those ages within the US Individuals with Disabilities Education Act (IDEA). Three years old is the age at which children in the United States qualify under the IDEA for an IEP, a customized plan that focuses on the educational needs of children in public school special education. Prior to age three, children qualify for an Individualized Family Service Plan, which revolves around the services that a family needs to support child development. At age fourteen, the IDEA requires that the IEP include a statement regarding the student's transition into the postschool world (though this regularly does not happen in practice).

The specification that all parents have a child who uses the iPad and Proloquo2Go as a communication system evolved out of the early stages of this research project. In the initial phase (September–December 2012), I observed children using a range of different dedicated and nondedicated high-tech AAC devices. Starting in January 2013, I chose to limit my observations to parents of children using an iPad equipped with Proloquo2Go. My focus on Proloquo2Go was unintentional and largely due to increasingly frequent opportunities in the field to observe children using that particular AAC system. Danny's dad, Peter, noted on this trend: "Proloquo has its dominance, A, because they were first, B, because it's easy, C, they're $100 cheaper [than other AAC apps]. When you're on all the forums and the websites and Facebook and everything, everybody says ... Proloquo has become shorthand for an AAC app on an iPad. ... Proloquo is ubiquitous."

Selecting Proloquo2Go (and not one of hundreds of other paid and free AAC apps, such as the aptly named Speak for Yourself and TouchChat) allowed for a larger participant pool. As I met different speech-language pathologists at various disability and assistive technology events and conferences, and told them about my research, many had strong feelings about Proloquo2Go, both positive and negative. I explained to them that my interest was not in measuring Proloquo2Go's effectiveness from a clinical standpoint but rather in theorizing its role in communication from a sociological perspective.

Recruitment

After receiving university Institutional Review Board approval, parent participants were recruited in two main ways. First, some parents were identified through the Rossmore Regional Center (a pseudonym), one of twenty-one nonprofit private corporations in the state of California that operate through state funding via the Department of Developmental Services under the Lanterman Developmental Disabilities Services Act of 1969.[4]

Besides funding set aside for employing regional center staff and the costs of operation, regional centers pay individuals, organizations, and vendors in the community that provide services to regional center consumers.[5] One vendor that Rossmore works in partnership with is the Evergreen Assistive Technology Hub (a pseudonym), a nonprofit organization that provides assistive technology services to people with disabilities in Southern California, including AAC assessments, loaner iPads for conducing AAC trials, and in-home AAC training sessions with parents.

Prior written permission was obtained through Rossmore for parents to be recruited. Potential Rossmore/Evergreen participants were approached with a recruitment document containing information about the study via an intermediary—their speech-language pathologist from Rossmore/Evergreen, Rachel or Caren—in order to reduce pressure to participate. If the potential participants agreed to take part in the study via phone or e-mail exchange with Rachel or Caren, then they notified me of upcoming iPad/Proloquo2Go training appointments at the participant's home. Eight families were recruited in this manner.

In order to access a larger subject pool of parents, I attempted to use snowball sampling, a recruitment method in which one participant suggests additional participants through their social network. I found the method to be difficult to implement for this study, though. Few parents with whom I spoke knew of other parents whose children were using Proloquo2Go on the iPad. Some did know of other parents (e.g., those of their children's classmates), but many felt as though they did not have a close enough relationship to those other parents to pass along information about my study. Parents' isolation from one another was reflected in my sampling challenges.

Ultimately, additional parent participants were located via referrals through various gatekeepers, including local speech-language pathologists, educators, and organizers of several nonprofit groups and e-mail lists in the Los Angeles area focused on supporting families of children with disabilities. Online forums present various benefits and drawbacks in recruiting participants for studies on technology use.[6] These parents are easier to reach, but there is a sampling bias inherent in recruiting parents through online means (e.g., Facebook groups) when making general claims about family technology use that applies to those across the socioeconomic spectrum. Those I found via the Internet, such as Chike, Peter, and Alice, were already more economically, socially, and culturally resourced. Twelve families were recruited through this imperfect referral strategy. Attention throughout the study was paid to reaching parents from a diverse range of socioeconomic, racial, and ethnic backgrounds, using what Thomas Lindlof and Bryan Taylor term "maximum variation sampling."[7]

None of the parents were economically compensated for their involvement in the research, and a number of them mentioned other motivations for participating. When I told Vanessa that I was grateful for her time, she responded, "That's one thing about being in the special needs community and population, is that parents are always sharing. We're always sharing with each other, like just last night on Facebook. I belong to some autism groups on Facebook. A mom in Kentucky reached out. She and I were messaging. It's so cool!" Vanessa likened her participation in my research to another form of sharing with the "special needs community and population." In turn, she highlighted my responsibility to give back to the community through my research.

One way in which I strived to do this was to share helpful technology tips that I learned from other parents in interviews or from Rachel and Caren in home training sessions. I passed along information regarding the brands of iPad cases that parents mentioned liking, strategies for going out to the movies with multiple children including one on the autism spectrum, how to integrate Prolqouo2Go with Dropbox as a way to back up the software, and how to do Google searches for other relevant apps. There was a need for this because comfort with technology was uneven among parents across social class, and as I mentioned earlier, because many of the parents I spoke with did not know other parents of children who communicated using the iPad with Proloquo2Go. In these small ways, I strived to redistribute among families the information to which I was privileged to gain access.

Study Design

The study employed two different qualitative methods: participant observation and semistructured in-depth interviews. Table M.3 details which families I observed and/or interviewed. The fieldwork was conducted over sixteen months, from October 2012 to April 2014. Prior to the first observation or interview, parents received an information sheet about the project and a child assent form. Parents consented to their and their children's participation.

Participant Observation

My primary sites of participant observation were family homes. Home-based participant observations were approximately ninety minutes in length. At each home observation (conducted October 2012 to September 2013), I shadowed either Rachel or Caren. Table M.4 denotes whose clients were whose. One or both parents were present for all observations when the child was present. Home observations were naturalistic in nature: others present sometimes included siblings, grandparents, household aids, and other therapists, such as applied behavior analysis therapists for autistic children.

The speech-language pathologists and families arranged the appointments independent of my schedule. I could only attend appointments that did not conflict with my other responsibilities, including taking classes, serving as a teaching assistant, and being a research assistant on faculty research projects. Frequent appointment cancellations by both speech-language pathologists and parents (due to issues such as illness, jury duty, miscommunication, and forgetfulness) required flexibility in my schedule. For instance, one time I drove over twenty miles to attend an appointment at Beatriz's apartment only to discover that her iPad had been left at school and the training session was canceled. In all, I conducted participant observations with eight families, each for one to three at-home technology training sessions.

In addition to homes, I tried to hang out with Rossmore parents and children by conducting observations at other sites. I observed three hour-long group meetings held at Rossmore for adolescents with developmental disabilities who use high-tech AAC systems. At each meeting, two to four young people were present along with one or two of their parents. I attended a ninety-minute parent training session for Proloquo2Go, also held at Rossmore. In both February 2013 and February 2014, I participated in an annual

Table M.3
Parents and Children Observed and/or Interviewed

Child	Parent/s observed/ interviewed	Observation and/or interview	Number of home observations
Paul Michael	Garine and Levon (mom and dad)	Observation	3
Beatriz	Pilar and David (mom and dad)	Observation and interview	2
Nash	Taylor and Todd (mom and dad)	Observation	2
James	Cathy (mom)*	Observation	1
Madeline	Teresa (mom)*	Observation	1
Thomas	Daisy (mom)*	Observation and interview	2
Stephanie	Marisa and Nelson (mom and dad)	Observation and interview	1
Pargev	Karun (mom)*	Observation and interview	2
Raul^	Nina (mom)*	Interview	N/A
Talen^	Kameelah (mom)*	Interview	N/A
Luke^	Debra and Rob (mom and dad)	Interview	N/A
Danny	Alice and Peter (mom and dad)	Interview	N/A
Isaac^	Sara (mom)*	Interview	N/A
Kevin^	Rebecca (mom)*	Interview	N/A
Eric	Anne (mom)*	Interview	N/A
Chike^	Esosa (mom)*	Interview	N/A
Cory^	Perri (mom)*	Interview	N/A
Sam	Donna (mom)*	Interview	N/A
River	Mark (dad)**	Interview	N/A
Moira^	Vanessa (mom)**	Interview	N/A

* Married parent, but only one parent was observed or interviewed.

** Single or divorced parent.

^ Child was present during the interview.

Table M.4
Child Clients of Rachel and Caren

Rachel	Caren
Paul Michael	Madeline
Beatriz	Thomas
Nash	Sam
James	
Stephanie	
Pargev	

daylong assistive technology conference organized by Evergreen. Parents (some of whom I has already observed and/or interviewed), educators, therapists, and assistive technology users attended the conference. The convening not only supplied me with up-to-date information about AAC devices (including presentations by adult AAC users) but an additional context to observe parents learning how to use the technology.

Beyond events hosted by Evergreen or Rossmore, I spent time at events in the Los Angeles area concerning AAC, assistive technology, and disability that were open to the public. In September 2013, I attended a gathering held at the University of Southern California to celebrate the launch of *Interacting with Autism*, an informative website on autism produced by university faculty and students. The event featured expert panels and activities for children. Attendees included local families, autistic individuals, health care providers, educators, and disability advocates. In spring 2012 and 2013, I also attended the Abilities Expo in downtown Los Angeles, where I gathered additional information about current trends in assistive technology development, and spoke with assistive technology users and experts.

Parent Interviews

I conducted seventeen semistructured in-depth interviews with parents. Fifteen of these were onetime interviews with families, and I conducted two interviews (an initial and a follow-up) with one family. Interviews generally took place at families' homes or at coffee shops. One parent, Donna, whom I had already met in person at the annual Evergreen assistive technology conference, asked for a remote interview by phone (instead of in person) because she was "dealing with some medical problems" with Sam. Later, Donna had to postpone a second time because Sam had been sent to the emergency room. I learned to expect the unexpected, and follow parents'

Table M.5
Selected Questions from Parent Interviews

Selected Questions from Parent Interviews
What is the "origin story" for your child's iPad with the Proloquo2Go app installed?
What do you like or dislike about the iPad? About Proloquo2Go?
What iPad apps does your child use at home, besides Proloquo2Go?
What other media and technology does your child use at home?
Have you ever sought out more information about the iPad? About Proloquo2Go? What resources have you found helpful?
Who does your child usually use media or technology with?
Do you have any advice for parents in a position similar to you?

lead in terms of how I could fit best into their schedules and routines. For example, Donna asked that I text her before I called so that the ring of the family's landline phone would not wake up a sleeping Sam.

Parent interviews lasted from one to two and a half hours. Parents were asked open-ended questions about the process by which they obtained their child's iPad, their experiences with the iPad thus far, and their engagement with other media and technology in the home as well as on the move (see table M.5). Following the interview, most parents received a family background information form to either fill out immediately or mail back to me in a self-addressed stamped envelope. I erred, though, in not also presenting this information sheet to families that I only observed and did not interview (four families in total), so demographic information among these parents is based solely on my observations along with follow-up conversations with Rachel and Caren.

Assistive Technology and AAC Professional Interviews

While I got to spend a great deal of time with Rachel and Caren, I wanted to speak directly with speech-language pathologists and assistive technology experts who were staff at school districts, not just Evergreen and Rossmore. I conducted additional interviews with six assistive technology professionals employed across two different school districts within the Los Angeles area. Each had knowledge of their district's deployment of the iPad and Proloquo2Go as assistive technologies. These interviews provided necessary background for understanding the role of school districts in shaping children's out-of-school use of the iPad and Proloquo2Go. I spoke with the assistive technology directors for two different school districts as well as

Table M.6

Selected Questions from Assistive Technology and AAC Professional Interviews

Within your school district, how does a parent obtain an iPad equipped with Proloquo2Go for use as an AAC device?
Do you notice any differences between parents who purchase their own iPad with Proloquo2Go for their child to use versus those who obtain it through the school district (e.g., in terms of their experiences working with the school or attitudes toward technology)?
What do you think are the pros and cons of a child using the iPad and/or Proloquo2Go for AAC compared to other dedicated AAC devices?
How do you think the process by which a child obtains an iPad as an AAC device through your school district could be improved?

four AAC specialists from one of those school districts, one of whom I interviewed twice.

These interviewees were asked questions regarding their district's processes and protocols for managing the use of iPads as AAC devices, their experiences working with parents, and challenges in the transfer of the child's iPad between home and school locales (see table M.6). Interviews ranged from forty to ninety minutes, and took place either at a coffee shop, in administrative offices, or on school grounds (before school, after school, and during lunch breaks).

Analysis

Interviews with parents as well as assistive technology and AAC professionals were audio recorded and transcribed verbatim, and analyzed along with comprehensive field notes compiled immediately after each interview and during each observation. I applied a constant comparative method approach to data analysis, coding throughout the course of fieldwork and paying close attention to patterns that I noticed emerging (such as distinctions among parents by social class).[8] In developing grounded theory, I employed open and selective coding and recoding of the data in order to identify concepts and categories.[9] Such an approach to qualitative data analysis allows for issues and topics to arise through natural conversation, as opposed to preordained variables imposed by the researcher. For instance, my interest in how parents' understood the cases on their child's iPad (see chapter 3)—an otherwise-invisible aspect of the technology—emerged from observations; as a result, I began to ask other parents about iPad cases in subsequent interviews.

Research Limitations

Although this book develops greater knowledge about how parents understand media and technology use among their children with communication disabilities, there are a number of limitations to the work. The first is the sample size of twenty families. It would be impossible to make general claims about the parents of all children with complex communication needs or developmental disabilities. This work does not reflect the full scope of children's activities with media and technology, nor parents' interpretations of these activities. This book, rather, details the process by which individual and collective uses of the iPad, as a form of AAC, shapes and is shaped by the social practices, spaces, and routines of families' everyday lives.

Second, my findings are primarily based on parent reports. I did directly observe children who use AAC as part of home observations and interviews, but did not conduct extensive ethnographic work in homes or do interviews directly with nonspeaking youth. I was only able to attend three monthly AAC user group meetings for tweens and teens at Rossmore over the span of sixteen months because they were regularly canceled due to low attendance. The tools I used to capture meaning—audio recorders and handwritten notes—also privileged speaking individuals. It was only after my research ended that I realized video recordings or data log files of words spoken through Proloquo2Go would have served as more easily reviewable documentation of communication by the child participants.

Third, another limitation is that among parents who agreed to participate in the study, there is a skew toward parents of children on the autism spectrum. This may explain the large number of boys in the study as well, as autism is underdiagnosed in the United States among girls.[10] Fourth, my participant observation work in homes is limited to shadowing speech-language pathologists, as opposed to another kind of therapist that might employ AAC in their work with children, such as an occupational or applied behavior analysis therapist. Fifth, my observations also did not extend into classrooms but rather focused on the ways in which parents discussed issues related to formal education at home.

Finally, perhaps the greatest shortcoming is that the insights that I present in this book are limited to the extent that I do not identify as having a disability, and am not the child of someone with a disability, the parent of a child with a disability, or a parent at all. I also identify as white, female, and upper middle class. In conducting this research, I held myself responsible to the parents who told me their stories—sensitive to the power inherent in interpreting and sharing their experiences through my analytic lens. I take seriously the task of drawing on people's life stories as raw material for a book and my accountability to them. Researchers raise the volume of unheard voices, but I make no claim to give voice to my participants or provide a holistic representation of their lives.

While the policies and perceptions that participants encountered may not directly affect me, all families are indirectly interconnected despite structures that systematically isolate and remove certain families from the center of society. Disability is at the heart of the human experience, and the study of disability, families, and communication technologies has important implications for humanity in its entirety.

Notes

Chapter 1

1. American Speech-Language-Hearing Association (ASHA), "Childhood Apraxia of Speech," n.d., http://www.asha.org/public/speech/disorders/ChildhoodApraxia/. Henceforth, I will use "children with disabilities" and "disabled children" interchangeably. I do this with the recognition that some individuals prefer the former and others the latter, and that this also varies by disability and cultural context. "Children with disabilities" is an example of "person-first" language for disability, while the latter is an example of "identity-first" language. In any case, all people should have the right to decide how they would like others to describe them, and the right to determine the best interests of their welfare. See Irving K. Zola, "Self, Identity, and the Naming Question: Reflections on the Language of Disability," *Social Science and Medicine* 36, no. 2 (1993): 167–173; Amartya Sen, *Development as Freedom* (New York: Anchor Books, 1999).

2. Emily B. Hager, "For Children Who Cannot Speak, a True Voice via Technology," *New York Times*, July 26, 2012, B3; Steven A. Rosenberg, "Tradition Gets an Update: Technology Helps Autistic 12-Year-Old Find a Voice for His Bar Mitzvah," *Boston Globe*, March 4, 2012, B1; Callie Carmichael, "How Tablets Helped Unlock One Girl's Voice," *CNN* (blog), November 14, 2012, http://www.cnn.com/2012/11/14/tech/mobile/carly-fleischmann-mobile-autism/.

3. *Wall Street Journal* (Unnamed staff reporter), "Concern Is Offering an Electronic Voice to Vocally Impaired," *Wall Street Journal*, November 2, 1977, 20.

4. Dennis McLellan, "Electronic Help for the Handicapped: The Voiceless Break Their Silence," *Los Angeles Times*, December 29, 1980, B1.

5. AssistiveWare, "Proloquo2Go Flyer," 2013, http://download.assistiveware.com/proloquo2go/files/Proloquo2Go_Flyer.pdf; AssistiveWare, "Proloquo2Go," n.d., http://www.assistiveware.com/product/proloquo2go.

6. Pippa Norris and Dieter Zinnbauer, "Giving Voice to the Voiceless: Good Governance, Human Development, and Mass Communications," Human Development

Report Office Occasional Paper 2002/2011 (New York: United Nations Development Program, 2002).

7. Sasha Costanza-Chock, *Out of the Shadows, Into the Streets: Transmedia Organizing and the Immigrant Rights Movement* (Cambridge, MA: MIT Press, 2014); Jonathan Donner, *After Access: Inclusion, Development, and a More Mobile Internet* (Cambridge, MA: MIT Press, 2015); Christina Dunbar-Hester, *Low Power to the People: Pirates, Protest, and Politics in FM Radio Activism* (Cambridge, MA: MIT Press, 2014).

8. Nick Couldry, "Voiceblind: Beyond the Paradoxes of the Neoliberal State," in *Reclaiming the Public Sphere: Communication, Power and Social Change*, ed. Tina Askanius and Liv S. Østergaard (New York: Palgrave Macmillan, 2014), 23.

9. James I. Charlton, *Nothing about Us without Us: Disability Oppression and Empowerment* (Berkeley: University of California Press, 2000). The saying, along with other strategic communication strategies and practices, eventually moved from the disability rights movements to other movements along with their struggles for social, cultural, and political justice. See Sasha Costanza-Chock, "Mic Check! Media Cultures and the Occupy Movement," *Social Movement Studies: Journal of Social, Cultural, and Political Protest* 11, no. 3–4 (2012): 1–11.

10. World Health Organization (WHO), *World Report on Disability* (Geneva: WHO Press, 2011).

11. Elizabeth Ellcessor, *Restricted Access: Media, Disability, and the Politics of Participation* (New York: NYU Press, 2016); Gerard Goggin and Christopher Newell, *Digital Disability: The Social Construction of Disability in New Media* (Lanham, MD: Rowman and Littlefield, 2003); Bill Kirkpatrick, "'A Blessed Boon': Radio, Disability, Governmentality, and the Discourse of the 'Shut-In,' 1920–1930," *Critical Studies in Media Communication* 29, no. 3 (2012): 165–184.

12. Meryl Alper, *Digital Youth with Disabilities* (Cambridge, MA: MIT Press, 2014); Julie P. Elman, *Chronic Youth: Disability, Sexuality, and US Media Cultures of Rehabilitation* (New York: NYU Press, 2014).

13. This book does not attempt or claim to clinically measure the effectiveness of Proloquo2Go as an AAC app.

14. Rachel and Caren as well as the names of all other research participants are pseudonyms.

15. Leah A. Lievrouw and Sonia Livingstone, introduction to *Handbook of New Media: Social Shaping and Social Consequences*, ed. Leah A. Lievrouw and Sonia Livingstone (London: Sage, 2006), 1–14.

16. Pierre Bourdieu, *Distinction: A Social Critique of the Judgment of Taste*, trans. Richard Nice (Cambridge, MA: Harvard University Press, 1984); Pierre Bourdieu, "The Forms of Capital," in *Handbook of Theory and Research for the Sociology of*

Education, ed. John G. Richardson (Westport, CT: Greenwood Press, 1986), 241–258; Annette Lareau, *Home Advantage: Social Class and Parental Intervention in Elementary Education* (Oxford: Rowman and Littlefield, 2000); Annette Lareau, *Unequal Childhoods: Class, Race, and Family Life* (Berkeley: University of California Press, 2003); Colin Ong-Dean, *Distinguishing Disability: Parents, Privilege, and Special Education* (Chicago: University of Chicago Press, 2009); Ellen Seiter, "Practicing at Home: Computers, Pianos, and Cultural Capital," in *Digital Youth, Innovation, and the Unexpected*, ed. Tara McPherson (Cambridge, MA: MIT Press, 2008), 27–52; Jonathan Sterne, "Bourdieu, Technique, and Technology," *Cultural Studies* 17, no. 3–4 (2003): 367–389; Audrey A. Trainor, "Reexamining the Promise of Parent Participation in Special Education: An Analysis of Cultural and Social Capital," *Anthropology and Education Quarterly* 41, no. 3 (2010): 245–263.

17. Sherry Turkle, *Reclaiming Conversation: The Power of Talk in a Digital Age* (New York: Penguin, 2015).

18. Haeyoun Park, "Paris Attacks Intensify Debate over How Many Syrian Refugees to Allow into the U.S," *New York Times*, November 19, 2015, http://www.nytimes.com/interactive/2015/10/21/us/where-syrian-refugees-are-in-the-united-states.html.

19. James P. Allen and Eugene Turner, "Ethnic Change and Enclaves in Los Angeles," *American Association of Geographers* (blog), March 8, 2013, http://www.aag.org/cs/news_detail?pressrelease.id=2058.

20. Steve Silberman, *NeuroTribes: The Legacy of Autism and the Future of Neurodiversity* (New York: Avery, 2015).

21. David Niemeijer, "The State of AAC in English-Speaking Countries: First Results from the Survey," *AssistiveWare* (blog), n.d., http://www.assistiveware.com/state-aac-english-speaking-countries-first-results-survey.

22. Seiter, "Practicing at Home." Mitchel Resnick, Amy Bruckman, and Fred Martin ("Pianos Not Stereos: Creating Computational Construction Kits," *Interactions* 3, no. 6 [1996]: 40–50) do this earlier, although through a different focus on human–computer interaction and constructivist learning.

23. Seiter, "Practicing at Home," 47.

24. Basil Bernstein, *Class, Codes, and Control*, 2nd ed., vol. 3 (London: Routledge and Kegan Paul, 1977).

25. David Buckingham and Margaret Scanlon, *Education, Entertainment, and Learning in the Home* (Philadelphia: Open University Press, 2003); Sonia Livingstone, *Young People and New Media: Childhood and the Changing Media Environment* (London: Sage, 2002); Ellen Seiter, *The Internet Playground: Children's Access, Entertainment, and Mis-Education* (New York: Peter Lang, 2005).

26. Seiter, "Practicing at Home," 28; Paul Attewell and Juan Battle, "Home Computers and School Performance," *Information Society* 15, no. 1 (1999): 1–10.

27. Mizuko Ito et al., *Connected Learning: An Agenda for Research and Design* (Irvine, CA: Digital Media and Learning Research Hub, 2013).

28. Bourdieu, *Distinction*; Bourdieu, "The Forms of Capital."

29. W.E.B. Du Bois, *Black Reconstruction in America, 1860–1880* (New York: Atheneum, 1969); Peggy McIntosh, "White Privilege and Male Privilege: A Personal Account of Coming to See Correspondences through Work in Women's Studies" (working paper, Wellesley College Center for Research on Women, Wellesley, MA, 1988).

30. W. Lloyd Warner, Marchia Meeker, and Kenneth Eells, *Social Class in America: A Manual of Procedure for the Measurement of Social Status* (Oxford, UK: Science Research Associates, 1949); Max Weber, *The Theory of Social and Economic Organization*, trans. Talcott Parsons (New York: Oxford University Press, 1947).

31. Michèle Lamont and Annette Lareau, "Cultural Capital: Allusions, Gaps, and Glissandos in Recent Theoretical Developments," *Sociological Theory* 6, no. 2 (1988): 153–168.

32. Barbara Rogoff, *The Cultural Nature of Human Development* (New York: Oxford University Press, 2003).

33. Lamont and Lareau, "Cultural Capital."

34. Lareau, *Home Advantage*; Lareau, *Unequal Childhoods*.

35. Lareau, *Home Advantage*.

36. Mario L. Small, *Unanticipated Gains: Origins of Network Inequality in Everyday Life* (Oxford: Oxford University Press, 2010).

37. Lareau, *Unequal Childhoods*.

38. Trainor, "Reexamining the Promise of Parent Participation"; Ong-Dean, *Distinguishing Disability*.

39. Trainor, "Reexamining the Promise of Parent Participation," 248.

40. US Department of Education, Institute of Education Sciences, National Center for Education Statistics, "The Condition of Education: Children and Youth with Disabilities," last updated May 2015, http://nces.ed.gov/programs/coe/indicator_cgg.asp.

41. Heather A. Horst, Becky Herr-Stephenson, and Laura Robinson, "Media Ecologies," in *Hanging Out, Messing Around, and Geeking Out: Kids Living and Learning with New Media*, ed. Mizuko Ito et al. (Cambridge, MA: MIT Press, 2010), 29–78; Seiter, *The Internet Playground*.

42. Bourdieu, "The Forms of Capital."

43. Stewart M. Hoover, Lynn Schofield Clark, and Diane Alters, *Media, Home, and Family* (New York: Routledge, 2004); Lynn Schofield Clark, *The Parent App: Understanding Families in the Digital Age* (New York: Oxford University Press, 2013); Vikki S. Katz, "How Children of Immigrants Use Media to Connect Their Families to the Community," *Journal of Children and Media* 4, no. 3 (2010): 298–315; Vikki S. Katz, *Kids in the Middle: How Children of Immigrants Negotiate Community Interactions for Their Families* (New Brunswick, NJ: Rutgers University Press, 2014).

44. Clark, *The Parent App*; Amy B. Jordan, "Social Class, Temporal Orientation, and Mass Media Use within the Family System," *Critical Studies in Media Communication* 9 (1992): 374–386; Ron Warren, "Parental Mediation of Children's Television Viewing in Low-Income Families," *Journal of Communication* 55, no. 4 (2005): 847–863; Allison J. Pugh, *Longing and Belonging: Parents, Children, and Consumer Culture* (Berkeley: University of California Press, 2009).

45. Bourdieu, "The Forms of Capital"; Susan Halford and Mike Savage, "Reconceptualizing Digital Social Inequality," *Information, Communication, and Society* 13, no. 7 (2010): 937–955.

46. Pierre Bourdieu and Loïc J. D. Wacquant, *An Invitation to Reflexive Sociology* (Chicago: University of Chicago Press, 1992), 93; Sterne, "Bourdieu, Technique, and Technology."

47. Lynette Kvasny, "Cultural (Re)production of Digital Inequality in a U.S. Community Technology Initiative," *Information, Communication, and Society* 9, no. 2 (2006): 160–181.

48. David Beukelman and Pat Mirenda, *Augmentative and Alternative Communication: Supporting Children and Adults with Complex Communication Needs*, 4th ed. (Baltimore: Brookes Publishing, 2013).

49. American Speech-Language-Hearing Association (ASHA), *Communication Facts: Special Populations: Augmentative and Alternative Communication, 2008 Edition* (Rockville, MD: ASHA, 2008).

50. National Academies of Sciences, Engineering, and Medicine, *Speech and Language Disorders in Children: Implications for the Social Security Administration's Supplemental Security Income Program* (Washington, DC: National Academies of Sciences, Engineering, and Medicine, 2016).

51. US Assistive Technology Act of 2004, Pub. L. No. 108–364 (2004).

52. US Individuals with Disabilities Education Act Amendments of 1997, Pub. L. No. 105–17 (1997).

53. World Health Organization (WHO), *International Classification of Functioning, Disability, and Health (ICF)* (Geneva: WHO Press, 2002).

54. Marshall McLuhan, *Understanding Media: The Extensions of Man,* reprint ed. (Cambridge, MA: MIT Press, 1994); Michel Foucault, "Technologies of the Self," in *Technologies of the Self: A Seminar with Michel Foucault,* ed. Huck Gutman, Patrick H. Hutton, and Luther H. Martin (Amherst: University of Massachusetts Press, 1988), 16–49; Donna J. Haraway, *Simians, Cyborgs, and Women: The Re-Invention of Nature* (New York: Routledge, 1991).

55. Donna J. Haraway, "Able Bodies and Companion Species," in *When Species Meet* (Minneapolis: University of Minnesota Press, 2007), 161–179; Don Ihde, "Embodying Hearing Devices: Digitalization," in *Listening and Voice: Phenomenologies of Sound,* 2nd ed. (Albany: State University of New York Press, 2007), 243–250.

56. Katherine Ott, "The Sum of Its Parts: An Introduction to Modern Histories of Prosthetics," in *Artificial Parts, Practical Lives: Modern Histories of Prosthetics,* ed. Katherine Ott, David Serlin, and Stephen Mihm (New York: NYU Press, 2002), 5.

57. Sara Hendren, "All Technology Is Assistive Technology," *Abler,* September 17, 2013, http://ablersite.org/2013/09/17/all-technology-is-assistive-technology/; Jennifer Mankoff, Gillian R. Hayes, and Devva Kasnitz, "Disability Studies as a Source of Critical Inquiry for the Field of Assistive Technology," in *Proceedings of the SIGACCESS Conference on Computers and Accessibility* (New York: ACM Press, 2010), 3–10; Graham Pullin, *Design Meets Disability* (Cambridge, MA: MIT Press, 2011).

58. Judith Newman, "To Siri with Love: How One Boy with Autism Became BFF with Apple's Siri," *New York Times,* October 17, 2014.

59. Ingunn Moser, "Disability and the Promises of Technology: Technology, Subjectivity, and Embodiment within an Order of the Normal," *Information, Communication, and Society* 9, no. 3 (2006): 373–395; Ingunn Moser and John Law, "'Making Voices': New Media Technologies, Disabilities, and Articulation," in *Digital Media Revisited: Theoretical and Conceptual Innovation in Digital Domains,* ed. Gunnar Liestøl, Terje Rasmussen, and Andrew Morrison (Cambridge, MA: MIT Press, 2003), 491–520; Joshua Reno, "Technically Speaking: On Equipping and Evaluating 'Unnatural' Language Learners," *American Anthropologist* 114, no. 3 (2012): 406–419.

60. Elizabeth Petrick, *Making Computers Accessible: Disability Rights and Digital Technology* (Baltimore: Johns Hopkins University Press, 2015); Bess Williamson, "Getting a Grip: Disability in American Industrial Design of the Late Twentieth Century," *Winterthur Portfolio* 46, no. 4 (2012): 213–236.

61. Marcia J. Scherer, *Living in the State of Stuck: How Assistive Technology Impacts the Lives of People with Disabilities,* 4th ed. (Cambridge, MA: Brookline Books, 2005).

62. Ott, "The Sum of Its Parts"; Kristen Shinohara and Jacob O. Wobbrock, "In the Shadow of Misperception: Assistive Technology Use and Social Interactions," in

Proceedings of the Annual Conference on Human Factors in Computing Systems (CHI) (New York: ACM Press, 2011), 705–714.

63. Jacquie Ripat and Roberta Woodgate, "The Intersection of Culture, Disability, and Assistive Technology," *Disability and Rehabilitation: Assistive Technology* 6, no. 2 (2011): 87–96.

64. Meryl Alper, "When Face-to-Face Is Screen-to-Screen: Reconsidering Mobile Media as Communication Augmentations and Alternatives," in *Routledge Companion to Disability and Media*, ed. Katie Ellis, Gerard Goggin, and Beth Haller (New York: Routledge, 2017).

65. Barb Dybwad, "2-Year-Old Finds iPad Easy to Use," *Mashable*, April 6, 2010, http://mashable.com/2010/04/06/2-year-old-girl-uses-ipad/.

66. Lynn Spigel, *Make Room for TV: Television and the Family Ideal in Postwar America* (Chicago: University of Chicago Press, 1992).

67. Raymond Williams, *Television: Technology and Cultural Form* (New York: Schocken Books, 1975); Lynn Spigel, "Media Homes: Then and Now," *International Journal of Cultural Studies* 4, no. 4 (2001): 385–411.

68. Clark, *The Parent App*; Hoover, Clark, and Alters, *Media, Home, and Family*; Horst, Herr-Stephenson, and Robinson, "Media Ecologies"; James Lull, *Inside Family Viewing: Ethnographic Research on Television's Audiences* (New York: Routledge, 1990); David Morley, *Family Television: Cultural Power and Domestic Leisure* (New York: Routledge, 1988); David Morley, *Home Territories: Media, Mobility, and Identity* (London: Routledge, 2000); Rivka Ribak, "Children and New Media: Reflections on the Ampersand," *Journal of Children and Media* 1, no. 1 (2007): 68–76; Spigel, *Make Room for TV*; Spigel, "Media Homes"; Leslie Haddon, "The Contribution of Domestication Research to In-Home Computing and Media Consumption," *Information Society* 22, no. 4 (2007): 195–203; Lori Reed, "Domesticating the Personal Computer: The Mainstreaming of a New Technology and the Cultural Management of a Widespread Technophobia, 1964–," *Critical Studies in Media Communication* 17, no. 2 (2000): 159–185; Roger Silverstone and Eric Hirsch, eds., *Consuming Technologies: Media and Information in Domestic Spaces* (London: Routledge, 1992).

69. Robert Rummel-Hudson, "A Revolution at Their Fingertips," *Perspectives on Augmentative and Alternative Communication* 20, no. 1 (2011): 19–23.

70. Dave Hershberger, "Mobile Technology and AAC Apps from an AAC Developer's Perspective," *Perspectives on Augmentative and Alternative Communication* 20, no. 1 (2011): 28–33; Jeff Higginbotham and Steve Jacobs, "The Future of the Android Operating System for Augmentative and Alternative Communication," *Perspectives on Augmentative and Alternative Communication* 20, no. 2 (2011): 52–56; David McNaughton and Janice Light, "The iPad and Mobile Technology Revolution: Benefits and Challenges for Individuals Who Require Augmentative and Alternative

Communication," *Augmentative and Alternative Communication* 29 (2013): 107–116; David Niemeijer, Anne M. Donnellan, and Jodi A. Robledo, *Taking the Pulse of Augmentative and Alternative Communication on iOS* (Amsterdam: AssistiveWare, 2012), http://www.assistiveware.com/taking-pulse-augmentative-and-alternative-communication-ios; Samuel Sennott, "An Introduction to the Special Issue on New Mobile AAC Technologies," *Perspectives on Augmentative and Alternative Communication* 20 (2011): 3–6.

71. Susan L. Parish et al., "State-Level Income Inequality and Family Burden of US Families Raising Children with Special Health Care Needs," *Social Science and Medicine* 74, no. 3 (2012): 399–407.

72. Greg Kumparak, "iTunes Rewind Highlights the Best Apps of 2009," *TechCrunch*, December 8, 2009, http://techcrunch.com/2009/12/08/itunes-rewind-highlights-the-best-apps-of-2009/.

73. Unlike many assistive technology and AAC system companies, AssistiveWare has made all subsequent updates to Proloquo2Go free of charge for those who have already purchased the app.

74. Todd Spangler, "Kids Love Apple's iPad More than Disney, Nickelodeon, YouTube: Survey," *Variety*, October 6, 2014, http://variety.com/2014/digital/news/kids-love-apples-ipad-more-than-disney-nickelodeon-youtube-survey-1201322460/.

75. Victoria Rideout, *Zero to Eight: Children's Media Use in America 2013* (San Francisco: Common Sense Media, 2013).

76. Ellen Wartella et al., *Parenting in the Age of Digital Technology: A National Survey* (Evanston, IL: Center on Media and Human Development, School of Communication, Northwestern University, 2013).

77. Petrick, *Making Computers Accessible.*

78. Apple, Inc., "Making a Difference. One App at a Time," iOS 7 promotional video, posted June 2013, https://www.youtube.com/watch?v=3k8hUtOmcaE.

79. Jean Burgess, "The iPhone Moment, the Apple Brand, and the Creative Consumer: From 'Hackability and Usability' to Cultural Generativity," in *Studying Mobile Media: Cultural Technologies, Mobile Communication, and the iPhone*, ed. Larissa Hjorth, Jean Burgess, and Ingrid Richardson (New York: Routledge, 2012), 28–42.

80. Michael W. Apple, *Teachers and Texts: A Political Economy of Class and Gender Relations in Education* (New York: Routledge, 1986); Michael W. Apple and Linda Christian-Smith, eds., *The Politics of the Textbook* (New York: Routledge, 1991); Larry Cuban, *Oversold and Underused: Computers in Classrooms, 1980–2000* (Cambridge, MA: Harvard University Press, 2001).

81. Dianne H. Angelo, "AAC in the Family and Home," in *The Handbook of Augmentative and Alternative Communication*, ed. Sharon Glennen and Denise C. DeCoste

(San Diego: Singular Publishing Group, 1997), 523–541; Janice Light and David McNaughton, "Supporting the Communication, Language, and Literacy Development of Children with Complex Communication Needs: State of the Science and Future Research Priorities," *Assistive Technology: The Official Journal of RESNA* 24, no. 1 (2012): 34–44.

82. Rita L. Bailey et al., "Family Members' Perceptions of Augmentative and Alternative Communication Device Use," *Language, Speech, and Hearing Services in Schools* 37 (2006): 50–60; Julie Marshall and Juliet Goldbart, "'Communication Is Everything I Think': Parenting a Child Who Needs Augmentative and Alternative Communication (AAC)," *International Journal of Language and Communication Disorders* 43, no. 1 (2008): 77–98; Shannon M. McCord and Gloria Soto, "Perceptions of AAC: An Ethnographic Investigation of Mexican-American Families," *Augmentative and Alternative Communication* 20, no. 4 (2004): 209–227; David McNaughton et al., "'A Child Needs to Be Given a Chance to Succeed': Parents of Individuals Who Use AAC Describe the Benefits and Challenges of Learning AAC Technologies," *Augmentative and Alternative Communication* 24, no. 1 (2008): 43–55; Gloria Soto, "Training Partners in AAC in Culturally Diverse Families," *Perspectives on Augmentative and Alternative Communication* 21 (2012): 144–150.

83. American Speech-Language-Hearing Association (ASHA), "Funding for Services," n.d., http://www.asha.org/NJC/Funding-for-Services/.

84. Ashlee Vance, "Insurers Fight Speech-Impairment Remedy," *New York Times*, September 14, 2009, http://www.nytimes.com/2009/09/15/technology/15speech.html; Lisa Satterfield, "The Latest on Medicare Coverage of SGDs," *ASHA Leader* 20 (2015): 28–29.

85. Steven C. White and Janet McCarty, "Reimbursement for AAC Devices," *ASHA Leader* 16 (2011): 3–7.

86. Laurence Wilson, Joel Kaiser, and Crystal Simpson, "Final Decision Memorandum," Centers for Medicare and Medicaid Services, July 29, 2015.

87. Ibid.

88. Allison M. Meder and Jane R. Wegner, "iPads, Mobile Technologies, and Communication Applications: A Survey of Family Wants, Needs, and Preferences," *Augmentative and Alternative Communication* 31, no. 1 (2015): 27–36.

89. Janice Light and David McNaughton, "Communicative Competence for Individuals Who Require Augmentative and Alternative Communication: A New Definition for a New Era of Communication?" *Augmentative and Alternative Communication* 30, no. 1 (2014): 1–18.

90. Niemeijer, Donnellan, and Robledo, *Taking the Pulse of Augmentative and Alternative Communication on iOS*.

91. Apple, Inc., "Volume Purchase Program: Proloquo2Go," n.d., https://volume.itunes.apple.com/us/app/proloquo2go-symbol-based-aac/id308368164?mt=8&term=Proloquo2Go&ign-mpt=uo%3D4.

92. Barbara Fernandes, "iTherapy: The Revolution of Mobile Devices within the Field of Speech Therapy," *Perspectives on School-Based Issues* 12, no. 2 (2011): 35–40.

93. Sharon Glennen and Denise C. DeCoste, eds., *The Handbook of Augmentative and Alternative Communication* (San Diego: Singular Publishing Group, 1997).

94. Henry Jenkins, "Introduction: 'Worship at the Altar of Convergence': A New Paradigm for Understanding Media Change," in *Convergence Culture: Where Old and New Media Collide* (New York: NYU Press, 2006), 1–24.

95. Susan M. Schweik, *The Ugly Laws: Disability in Public* (New York: NYU Press, 2009).

96. Silberman, *NeuroTribes*.

97. Sandra L. Friedman and Miriam A. Kalichman, "Out-of-Home Placement for Children and Adolescents with Disabilities," *Pediatrics* 134, no. 4 (2014): 836–846.

98. Jordynn Jack, *Autism and Gender: From Refrigerator Mothers to Computer Geeks* (Champaign: University of Illinois Press, 2014); Gail Landsman, *Reconstructing Motherhood and Disability in the Age of "Perfect" Babies* (New York: Routledge, 2009).

99. Rachel Adams, *Raising Henry: A Memoir of Motherhood, Disability, and Discovery* (New Haven, CT: Yale University Press, 2013); Linda Blum, *Raising Generation Rx: Mothering Kids with Invisible Disabilities in an Age of Inequality* (New York: NYU Press, 2015); Mary Langan, "Parental Voices and Controversies in Autism," *Disability and Society* 26, no. 2 (2011): 193–205; Sara Ryan and Katherine Runswick-Cole, "Repositioning Mothers: Mothers, Disabled Children, and Disability Studies," *Disability and Society* 23, no. 3 (2008): 199–210; Amy C. Sousa, "From Refrigerator Mothers to Warrior-Heroes: The Cultural Identity Transformation of Mothers Raising Children with Intellectual Disabilities," *Symbolic Interaction* 34, no. 2 (2011): 220–243.

100. Leo Kanner, "Problems of Nosology and Psychodynamics of Early Infantile Autism," *American Journal of Orthopsychiatry* 19, no. 3 (1949): 416–426; Bruno Bettelheim, *The Empty Fortress: Infantile Autism and the Birth of the Self* (New York: Free Press, 1967).

101. Bruno Bettelheim, "Joey: A 'Mechanical Boy,'" *Scientific American* 200 (1959): 116–127.

102. Silberman, *NeuroTribes*.

103. Liza Gross, "A Broken Trust: Lessons from the Vaccine-Autism Wars," *PLoS Biology* 7, no. 5 (2009), http://journals.plos.org/plosbiology/article?id=10.1371/journal.pbio.1000114.

104. Sharon Hays, *The Cultural Contradictions of Motherhood* (New Haven, CT: Yale University Press, 1996).

105. Sousa, "From Refrigerator Mothers to Warrior-Heroes," 221.

106. Lucinda P. Bernheimer, Ronald Gallimore, and Thomas S. Weisner, "Ecocultural Theory as a Context for the Individual Family Service Plan," *Journal of Early Intervention* 14, no. 3 (1990): 219–223; Joan M. Patterson, "Integrating Family Resilience and Family Stress Theory," *Journal of Marriage and the Family* 64, no. 2 (2002): 349–360.

107. Christine A. Maul and George H. S. Singer, "'Just Good Different Things': Specific Accommodations Families Make to Positively Adapt to Their Children with Developmental Disabilities," *Topics in Early Childhood Special Education* 29, no. 3 (2009): 155–170; Brenda Nally, Bob Houlton, and Sue Ralph, "Researches in Brief: The Management of Television and Video by Parents of Children with Autism," *Autism* 4, no. 3 (2000): 331–338; Howard C. Shane and Patti D. Albert, "Electronic Screen Media for Persons with Autism Spectrum Disorders: Results of a Survey," *Journal of Autism and Developmental Disorders* 38, no. 8 (2008): 1499–1508.

108. Bonnie Keilty and Kristin M. Galvin, "Physical and Social Adaptations of Families to Promote Learning in Everyday Experiences," *Topics in Early Childhood Special Education* 26, no. 4 (2006): 219–233.

109. Sara E. Green, "'We're Tired, Not Sad': Benefits and Burdens of Mothering a Child with a Disability," *Social Science and Medicine* 64, no. 1 (2007): 150–163.

110. Alison Kafer, *Feminist, Queer, Crip* (Bloomington: Indiana University Press, 2013); Robert McRuer, *Crip Theory: Cultural Signs of Queerness and Disability* (New York: NYU Press, 2006); Tobin Siebers, *Disability Theory* (Ann Arbor: University of Michigan Press, 2008); Maria Bakardjieva and Richard Smith, "The Internet in Everyday Life: Computer Networking from the Standpoint of the Domestic User," *New Media and Society* 3, no. 1 (2001): 67–83; Ellcessor, *Restricted Access*; Goggin and Newell, *Digital Disability*; Jonathan Sterne, *The Audible Past: Cultural Origins of Sound Reproduction* (Durham, NC: Duke University Press, 2003); Mara Mills, "When Mobile Communication Technologies Were New," *Endeavor* 33, no. 4 (2009): 140–146; Mara Mills, "Media and Prosthesis: The Vocoder, the Artificial Larynx, and the History of Signal Processing," *Qui Parle: Critical Humanities and Social Sciences* 21, no. 1 (2012): 107–149.

111. Joseph P. Shapiro, *No Pity: People with Disabilities Forging a New Civil Rights Movement* (New York: Times Books, 1993).

112. Lennard J. Davis, *Bending Over Backwards: Disability, Dismodernism, and Other Difficult Positions* (New York: NYU Press, 2002); Simi Linton, *Claiming Disability: Knowledge and Identity* (New York: NYU Press, 1998).

113. Michael Oliver, *The Politics of Disablement* (Basingstoke, UK: Macmillan, 1990); Tom Shakespeare, *Disability Rights and Wrongs Revisited* (New York: Routledge, 2013).

114. Vivian Sobchack, "A Leg to Stand On: Prosthetics, Metaphor, and Materiality," in *Carnal Thoughts: Embodiment and Moving Image Culture* (Berkeley: University of California Press, 2004), 205–225; Judith Butler, *Gender Trouble: Feminism and the Subversion of Identity* (New York: Routledge, 1990). Butler has also drawn critique from disability scholars for not addressing disability in discussions of bodily difference. See Ellen Samuels, "Critical Divides: Judith Butler's Body Theory and the Question of Disability," *NWSA Journal* 14, no. 3 (2002): 58–76.

115. Rosemarie Garland-Thomson, *Extraordinary Bodies: Figuring Physical Disability in American Culture and Literature* (New York: Columbia University Press, 1996), 7.

116. Kafer, *Feminist, Queer, Crip*, 26, 11.

117. Kerry Dobransky and Eszter Hargittai, "The Disability Divide in Internet Access and Use," *Information, Communication, and Society* 9, no. 3 (2006): 313–334; Mark Warschauer, *Technology and Social Inclusion: Rethinking the Digital Divide* (Cambridge, MA: MIT Press, 2003).

118. Throughout this book, the terms *Deaf* and *deaf* each appear. The former often refers to cultural and linguistic identification, and the latter to the medical condition of deafness. While these distinctions are not always useful or clear, I employ them differently based on knowledge of the individuals under discussion and those I quote directly.

Ellcessor, *Restricted Access*; Katie Ellis and Mike Kent, *Disability and New Media* (New York: Routledge, 2010); Gerard Goggin and Christopher Newell, "Disabling Cell Phones," in *The Cell Phone Reader*, ed. Anandam Kavoori and Noah Arceneaux (New York: Peter Lang, 2006), 155–172.

119. Goggin and Newell, *Digital Disability*.

120. Williams, *Television*.

121. Henry Jenkins et al., *Confronting the Challenges of Participatory Culture: Media Education for the 21st Century* (Chicago: John D. and Catherine T. MacArthur Foundation, 2006); Kylie A. Peppler and Mark Warschauer, "Uncovering Literacies, Disrupting Stereotypes: Examining the (Dis)abilities of a Child Learning to Computer Program and Read," *International Journal of Learning and Media* 3, no. 3 (2012): 15–41; Sylvia Söderström, "Offline Social Ties and Online Use of Computers: A Study of Disabled Youth and Their Use of ICT Advances," *New Media and Society* 11, no. 5 (2009): 709–727.

122. Melissa Dawe, "Desperately Seeking Simplicity: How Young Adults with Cognitive Disabilities and Their Families Adopt Assistive Technologies." In *Proceedings of the Annual Conference on Human Factors in Computing Systems (CHI)* (New York: ACM Press, 2006), 1143–1152.

123. Sonia Livingstone and Ellen Helsper, "Gradations in Digital Inclusion: Children, Young People, and the Digital Divide," *New Media and Society* 9, no. 4 (2007): 671–696; Anna Everett, ed., *Learning Race and Ethnicity: Youth and Digital Media* (Cambridge, MA: MIT Press, 2008); Mary L. Gray, *Out in the Country: Youth, Media, and Queer Visibility in Rural America* (New York: NYU Press, 2009); Gerard Goggin, "Youth Culture and Mobiles," *Mobile Media and Communication* 1, no. 1 (2013): 86.

124. Gail Landsman, "'Real Motherhood,' Class, and Children with Disabilities," in *Ideologies and Technologies of Motherhood: Race, Class, Sexuality, Nationalism*, ed. Heléna Ragoné and France W. Twine (New York: Routledge, 2000), 184, 185.

125. Amy J. Houtrow et al., "Changing Trends of Childhood Disability, 2001–2011," *Pediatrics* 134 (2014): 530–538.

126. Coleen A. Boyle et al., "Trends in the Prevalence of Developmental Disabilities in U.S. Children, 1997–2008," *Pediatrics* 127 (2011): 1034–1042.

127. Sue C. Lin, Stella M. Yu, and Robin L. Harwood, "Autism Spectrum Disorders and Developmental Disabilities in Children from Immigrant Families in the United States," *Pediatrics* 13, supplement 2 (2012): S191–S197.

128. Meryl Alper, Vikki S. Katz, and Lynn Schofield Clark, "Researching Children, Intersectionality, and Diversity in the Digital Age," *Journal of Children and Media* 10, no. 1 (2016): 107–114; Blum, *Raising Generation Rx*.

129. Kimberlé Crenshaw, "Demarginalizing the Intersection of Race and Sex: A Black Feminist Critique of Antidiscrimination Doctrine, Feminist Theory, and Antiracist Politics," *University of Chicago Legal Forum* 1 (1989): 139–167; Kimberlé Crenshaw, "Mapping the Margins: Intersectionality, Identity Politics, and Violence against Women of Color," *Stanford Law Review* 43, no. 6 (1991): 1241–1299.

130. Leslie McCall, "The Complexity of Intersectionality," *Signs* 30, no. 3 (2005): 1771–1800; Jennifer C. Nash, "Re-Thinking Intersectionality," *Feminist Review* 89 (2008): 1–15; Virginia Olesen, "Feminist Qualitative Research in the Millennium's First Decade," in *The Sage Handbook of Qualitative Research*, ed. Norman K. Denzin and Yvonna S. Lincoln (Thousand Oaks, CA: Sage, 2011), 129–146; Patricia H. Collins, *Fighting Words: Black Women and the Search for Justice* (Minneapolis: University of Minneapolis Press, 1998); Crenshaw, "Mapping the Margins."

131. Ange-Marie Hancock, "When Multiplication Doesn't Equal Quick Addition: Examining Intersectionality as a Research Paradigm," *Perspectives on Politics* 5, no. 1 (2007): 63–79; Sharon Hays, "Structure and Agency and the Sticky Problem of Culture," *Sociological Theory* 12, no. 1 (1994): 57–72.

132. Patricia H. Collins, "Reply to Commentaries. *Black Sexual Politics* Revisited," *Studies in Gender and Sexuality* 9, no. 1 (2008): 68–85.

133. Only the parents I observed or interviewed are listed. I did not spend any time, for instance, with Cathy's or Sara's partners, and so they are not named here.

134. The five-year (2009–2013) estimate of median household incomes for each county under study are $55,909 (Los Angeles County), $75,422 (Orange County), and $76,544 (Ventura County). See US Census Bureau, *QuickFacts: California*, n.d. http://quickfacts.census.gov/qfd/states/06000.html.

135. For an explanation of the regional center system in the state of California, see the chapter on methods.

136. Beth Haller et al., "iTechnology as Cure or iTechnology as Empowerment: What Do North American News Media Report?" *Disability Studies Quarterly* 36, no. 1 (2016), http://dsq-sds.org/article/view/3857/4208.

137. Beth Haller, *Representing Disability in an Ableist World: Essays on Mass Media* (Louisville, KY: Advocado Press, 2010).

138. Microsoft, "Microsoft 2014 Super Bowl Commercial: Empowering," YouTube video, posted February 2, 2014, https://www.youtube.com/watch?v=qaOvHKG0Tio.

139. Society of Professional Journalists, *SJP Code of Ethics* (Indianapolis, IN: Society of Professional Journalists, 2014), http://www.spj.org/ethicscode.asp.

140. Sarah Banet-Weiser and Kate Miltner, "#MasculinitySoFragile: Culture, Structure, and Networked Misogyny," *Feminist Media Studies* 16, no. 1 (2016): 171–174.

141. Nick Couldry, *Why Voice Matters: Culture and Politics After Neoliberalism* (London: Sage, 2010).

142. Jo Tacchi, "Digital Engagement: Voice and Participation in Development," in *Digital Anthropology*, ed. Heather A. Horst and Daniel Miller (New York: Berg, 2012), 225–241.

143. Victoria Rideout and Vikki S. Katz, *Opportunity for All? Technology and Learning in Lower-Income Families* (New York: Joan Ganz Cooney Center at Sesame Workshop, 2016).

144. Lisa Cartwright, *Moral Spectatorship: Technologies of Voice and Affect in Postwar Representations of the Child* (Durham, NC: Duke University Press, 2008), 161.

145. Kathy L. Look Howery, "Speech-Generating Devices in the Lives of Young People with Severe Speech Impairment: What Does the Non-Speaking Child Say?" in *Efficacy of Assistive Technology Interventions*, ed. Dave L. Edyburn (Bingley, UK: Emerald, 2015), 96.

146. E. Kay M. Tisdall, "The Challenge and Challenging of Childhood Studies? Learning from Disability Studies and Research with Disabled Children," *Children and Society* 26 (2012): 181–191.

147. Hays, "Structure and Agency."

148. Lisa Gitelman, "Introduction: Media as Historical Subjects," in *Always Already New: Media, History, and the Data of Culture* (Cambridge, MA: MIT Press, 2006), 7.

149. Don Norman, *The Design of Everyday Things* (New York: Doubleday, 1988).

150. Arjun Appadurai, ed., "Introduction: Commodities and the Politics of Value," in *The Social Life of Things: Commodities in Cultural Perspective* (Cambridge: Cambridge University Press, 1986), 3–63; Peter Nagy and Gina Neff, "Imagined Affordance: Reconstructing a Keyword for Communication Theory," *Social Media + Society* 1, no. 2 (2015): 1–9.

151. Faye Ginsburg, "Disability in the Digital Age," in *Digital Anthropology*, ed. Heather A. Horst and Daniel Miller (London: Berg, 2012), 101–126.

152. Sherry Turkle, *Alone Together: Why We Expect More from Technology and Less from Each Other* (New York: Basic Books, 2011); Turkle, *Reclaiming Conversation.*

153. Catherine Steiner-Adair, *The Big Disconnect: Protecting Childhood and Family Relationships in the Digital Age* (New York: HarperCollins, 2014), 62.

154. Turkle, *Reclaiming Conversation.*

155. Nancy Baym, *Personal Connections in the Digital Age*, 2nd ed. (Cambridge, UK: Polity Press, 2015); danah boyd, *It's Complicated: The Social Lives of Networked Teens* (New Haven, CT: Yale University Press, 2014).

Chapter 2

1. Nick Couldry, "Rethinking the Politics of Voice," *Continuum: Journal of Media and Cultural Studies* 23, no. 4 (2009): 579–582; Nick Couldry, *Why Voice Matters: Culture and Politics after Neoliberalism* (London: Sage, 2010); Mladen Dolar, *A Voice and Nothing More* (Cambridge, MA: MIT Press, 2006); Gerard Goggin, "Making Voice Portable: The Early History of the Cell Phone," in *Cell Phone Culture: Mobile Technology in Everyday Life* (New York: Routledge, 2006), 19–40; Ananda Mitra and Eric Watts, "Theorizing Cyberspace: The Idea of Voice Applied to the Internet Discourse," *New Media and Society* 4, no. 4 (2002): 479–498; John Durham Peters, *Speaking into the Air: A History of the Idea of Communication* (Chicago: University of Chicago Press, 1999); John Durham Peters, "The Voice between Phenomenology, Media, and Religion," *Glimpse: The Journal of the Society for Phenomenology and Media* 6 (2005): 1–10.

2. Allison James, Chris Jenks, and Alan Prout, *Theorizing Childhood* (Oxford: Polity Press, 1998); Sirkka Komulainen, "The Ambiguity of the Child's 'Voice' in Social Research," *Childhood* 14, no. 1 (2007): 11–28; Mary Wickenden, "Talking to Teenagers: Using Anthropological Methods to Explore Identity and the Lifeworlds of Young People Who Use AAC," *Communication Disorders Quarterly* 32, no. 3 (2011): 151–163.

3. Tanja Dreher, "Listening across Difference: Media and Multiculturalism beyond the Politics of Voice," *Journal of Media and Cultural Studies* 23, no. 4 (2009): 445–458; Tanja Dreher, "A Partial Promise of Voice: Digital Storytelling and the Limit of Listening," *Media International Australia* 142, no. 1 (2012): 157–166.

4. Joshua Reno, "Technically Speaking: On Equipping and Evaluating 'Unnatural' Language Learners," *American Anthropologist* 114, no. 3 (2012): 406–419.

5. James L. Flanagan, "Voices of Men and Machines," *Journal of the Acoustical Society of America* 51, no. 5 (1972): 1375–1387.

6. Theo van Leeuwen, "Vox Humana: The Instrumental Representation of the Human Voice," in *Voice: Vocal Aesthetics in Digital Arts and Media*, ed. Norie Neumark, Ross Gibson, and Theo van Leeuwen (Cambridge, MA: MIT Press, 2010), 5–15.

7. Dolar, *A Voice and Nothing More*; Thomas Hankins and Robert Silverman, "*Vox Mechanica*: The History of Speaking Machines," in *Instruments and the Imagination* (Princeton, NJ: Princeton University Press, 1995), 178–220.

8. David Beukelman and Pat Mirenda, *Augmentative and Alternative Communication: Supporting Children and Adults with Complex Communication Needs*, 4th ed. (Baltimore: Brookes Publishing, 2013).

9. Dennis H. Klatt, "Review of Text-to-Speech Conversion for English," *Journal of the Acoustical Society of America* 82, no. 3 (1987): 737–793.

10. Ben Gold, Nelson Morgan, and Dan Ellis, *Speech and Audio Signal Processing: Processing and Perception of Speech and Music*, 2nd ed. (Hoboken, NJ: Wiley, 2011).

11. David Lindsay, "Talking Head," *American Heritage of Invention and Technology* 13, no. 1 (1997): 57–63; Mara Mills, "Media and Prosthesis: The Vocoder, the Artificial Larynx, and the History of Signal Processing," *Qui Parle: Critical Humanities and Social Sciences* 21, no. 1 (2012): 107–149; Joseph P. Olive, "'The Talking Computer': Text to Speech Synthesis," in *HAL's Legacy: 2001's Computer as Dream and Reality*, ed. David Stork (Cambridge, MA: MIT Press, 1997), 101–129.

12. Roger Ebert, "Remaking My Voice," filmed March 2011, TED video, 19:29, http://www.ted.com/talks/roger_ebert_remaking_my_voice?language=en.

13. Spike Jonze, *Her* (Burbank, CA: Warner Bros., 2013).

14. Joanne Lasker and Jan Bedrosian, "Promoting Acceptance of Augmentative and Alternative Communication by Adults with Acquired Communication Disorders," *Augmentative and Alternative Communication* 17, no. 3 (2011): 141–153.

15. Martin Kevorkian, "Integrated Circuits," in *Color Monitors: The Black Face of Technology in America* (Ithaca, NY: Cornell University Press, 2006), 74–114.

16. Aviva Rutkin, "A Voice to Call Your Own," *New Scientist* 221, no. 2961 (2014): 22.

17. Sarah Creer, Stuart Cunningham, Phil Green, and Yamagishi Junichi, "Building Personalised Synthetic Voices for Individuals with Severe Speech Impairment," *Computer Speech and Language* 27, no. 6 (2013): 1178–1193.

18. Hélène Mialet, *Hawking Incorporated: Stephen Hawking and the Anthropology of the Knowing Subject* (Chicago: University of Chicago Press, 2012).

19. AssistiveWare, *Proloquo2Go Manual Version 4.1* (Amsterdam: AssistiveWare, 2015), http://download.assistiveware.com/proloquo2go/helpfiles/4.1/en/assets/Proloquo2GoManual.pdf.

20. AssistiveWare, "Voices: U.S. English," n.d., http://www.assistiveware.com/product/proloquo2go/voices#en_US.

21. AssistiveWare, "AssistiveWare—The Making of the American Children's Voices Josh and Ella," YouTube video, posted July 25, 2012, 1:45, https://www.youtube.com/watch?v=OsI17MZIq0U.

22. James E. Katz and Mark Aakhus, eds., *Perpetual Contact: Mobile Communication, Private Talk, Public Performance* (Cambridge: Cambridge University Press, 2002); Peters, *Speaking into the Air.*

23. Jacques Derrida, *Voice and Phenomenon: Introduction to the Problem of the Sign in Husserl's Phenomenology*, trans. Leonard Lawlor (Evanston, IL: Northwestern University Press, 2011); Walter J. Ong, *Orality and Literacy: The Technologizing of the Word* (New York: Methuen, 1982).

24. Walter J. Ong, *Interfaces of the Word* (Ithaca, NY: Cornell University Press, 1977).

25. Brian. K. Axel, "Anthropology and the New Technologies of Communication," *Cultural Anthropology* 21, no. 3 (2006): 354–384; Mills, "Media and Prosthesis."

26. N. Katherine Hayles, *How We Became Posthuman: Virtual Bodies in Cybernetics, Literature, and Informatics* (Chicago: University of Chicago Press, 1999).

27. Gerard Goggin, "Disability and the Ethics of Listening," *Continuum: Journal of Media and Cultural Studies* 23, no. 4 (2009): 489–502.

28. Christine Ashby, "Whose 'Voice' Is It Anyway?: Giving Voice and Qualitative Research Involving Individuals That Type to Communicate," *Disability Studies Quarterly* 31, no. 4 (2011), http://dsq-sds.org/article/view/1723/1771; Komulainen, "The Ambiguity of the Child's 'Voice' in Social Research"; Ingunn Moser and John Law, "'Making Voices': New Media Technologies, Disabilities, and Articulation," in *Digital Media Revisited: Theoretical and Conceptual Innovation in Digital Domains*, ed. Gunnar Liestøl, Terje Rasmussen, and Andrew Morrison (Cambridge, MA: MIT Press, 2003), 491–520.

29. Susan Bickford, *The Dissonance of Democracy: Listening, Conflict, and Citizenship* (Ithaca, NY: Cornell University Press, 1996); Jean Burgess, "Hearing Ordinary Voices: Cultural Studies, Vernacular Creativity, and Digital Storytelling," *Continuum: Journal of Media and Cultural Studies* 20, no. 2 (2006): 201–214; Couldry, *Why Voice Matters*; Kate Crawford, "Four Ways of Listening to an iPhone: From Sound and Network Listening to Biometric Data and Geolocative Tracking," in *Studying Mobile Media: Cultural Technologies, Mobile Communication, and the iPhone*, ed. Larissa Hjorth, Jean Burgess, and Ingrid Richardson (New York: Routledge, 2012), 213–239; Andrew Dobson, "Democracy and Nature: Speaking and Listening," *Political Studies* 58, no. 4 (2010): 752–768; Jim Macnamara, "Beyond Voice: Audience-Making and the Work and Architecture of Listening as New Media Literacies," *Continuum: Journal of Media and Cultural Studies* 27, no. 1 (2013): 160–175; Kate Lacey, *Listening Publics: The Politics and Experience of Listening in the Media Age* (Cambridge, UK: Polity Press, 2013).

30. Couldry, *Why Voice Matters.*

31. Dreher, "A Partial Promise of Voice," 159.

32. Mialet, *Hawking Incorporated*; Moser and Law, "Making Voices"; Srikala Naraian, "Disentangling the Social Threads within a Communicative Environment: A Cacophonous Tale of Alternative and Augmentative Communication (AAC)," *European Journal of Special Needs Education* 25, no. 3 (2010): 253–267.

33. Komulainen, "The Ambiguity of the Child's 'Voice' in Social Research."

34. Faye Ginsburg and Rayna Rapp, "Making Disability Count: Demography, Futurity, and the Making of Disability Publics," *Somatosphere*, May 11, 2015, http://somatosphere.net/2015/05/making-disability-count-demography-futurity-and-the-making-of-disability-publics.html; Goggin, "Disability and the Ethics of Listening."

35. Marshall McLuhan, *Understanding Media: The Extensions of Man*, reprint ed. (Cambridge, MA: MIT Press, 1994).

36. Leopoldina Fortunati, "Real People, Artificial Bodies," in *Mediating the Human Body: Technology, Communication, and Fashion*, ed. Leopoldina Fortunati, James E. Katz, and Raimonda Riccini (Mahwah, NJ: Lawrence Erlbaum, 2003), 62.

37. David McNaughton and Janice Light, "The iPad and Mobile Technology Revolution: Benefits and Challenges for Individuals Who Require Augmentative and Alternative Communication," *Augmentative and Alternative Communication* 29 (2013): 107–116.

38. Arlie R. Hochschild, *The Second Shift: Working Parents and the Revolution at Home* (Berkeley: University of California Press, 1989).

39. Lynn Schofield Clark, "Parental Mediation Theory for the Digital Age," *Communication Theory* 21, no. 4 (2011): 323–343; Ellen Seiter, *Sold Separately: Children and Parents in Consumer Culture* (New Brunswick, NJ: Rutgers University Press, 1995).

40. Connie Kasari et al., "Communication Interventions for Minimally Verbal Children with Autism: A Sequential Multiple Assignment Randomized Trial," *Journal of the American Academy of Child and Adolescent Psychiatry* 53, no. 6 (2014): 635–646; Mandy J. Rispoli et al., "The Use of Speech Generating Devices in Communication Interventions for Individuals with Developmental Disabilities: A Review of the Literature," *Developmental Neurorehabilitation* 13, no. 4 (2010): 276–293.

41. Ong, *Orality and Literacy*; Peters, *Speaking into the Air*; Peters, "The Voice between Phenomenology, Media, and Religion."

42. Lucy Diep and Gregor Wolbring, "Perceptions of Brain–Machine Interface Technology among Mothers of Disabled Children," *Disability Studies Quarterly* 35, no. 4 (2015), http://dsq-sds.org/article/view/3856/4112.

43. Silvia Lovato and Anne Marie Piper, "'Siri, Is This You?': Understanding Young Children's Interactions with Voice Input Systems," in *Proceedings of the Annual Conference on Interaction Design and Children (IDC)* (New York: ACM Press, 2015), 335–338.

44. Elizabeth J. Carter et al., "Are Children with Autism More Responsive to Animated Characters? A Study of Interactions with Humans and Human-Controlled Avatars," *Journal of Autism and Developmental Disorders* 44, no. 10 (2014): 2475–2485.

45. Tarleton Gillespie, "The Relevance of Algorithms," in *Media Technologies: Essays on Communication, Materiality, and Society*, ed. Tarleton Gillespie, Pablo J. Boczkowski, and Kirsten A. Foot (Cambridge, MA: MIT Press, 2014), 167–194.

46. Judith Newman, "To Siri with Love: How One Boy with Autism Became BFF with Apple's Siri," *New York Times*, October 17, 2014, http://www.nytimes.com/2014/10/19/fashion/how-apples-siri-became-one-autistic-boys-bff.html.

47. Joshua Reeves, "Automatic for the People: The Automation of Communicative Labor," *Communication and Critical/Cultural Studies* (forthcoming).

48. Erving Goffman, *The Presentation of Self in Everyday Life* (New York: Anchor Books, 1959).

49. AssistiveWare, "World's First Bilingual Spanish–English Children's Text to Speech Voices," posted January 28, 2014, http://www.assistiveware.com/worlds-first-bilingual-spanish-english-childrens-text-speech-voices.

50. AssistiveWare, "Award-Winning Communication App Now Gives a Voice to Spanish Children," posted September 30, 2015, https://www.assistiveware.com/award-winning-communication-app-now-gives-voice-spanish-children.

51. Graham Pullin, "Expression Meets Information," in *Design Meets Disability* (Cambridge, MA: MIT Press, 2011), 176.

52. Please note that all personally identifying information has been removed. This is not Moira and Vanessa's real phone number or area code.

53. AssistiveWare, *Proloquo2Go Manual Version 4.1.*

54. AssistiveWare, "Working with Recents," 2013, http://download.assistiveware.com/proloquo2go/files/version3/Proloquo2Go_3_Tutorial-Working_with_Recents.pdf.

55. Michael B. Williams, "Privacy and AAC," *Alternatively Speaking* 5, no. 2 (2000): 1–5.

56. Katya Hill, Barry Romich, and Snoopi J. Botten, "Rights and Privacy in AAC Evidence-Based Clinical Practice," in *Proceedings of the Technology and Persons with Disabilities Conference* (Northridge: California State University, Northridge, 2002), http://www.csun.edu/~hfdss006/conf/2002/proceedings/247.htm.

57. Jeff Higginbotham, "Problems and Solutions in Data Logging," *Alternatively Speaking* 5, no. 2 (2000): 4–5; Williams, "Privacy and AAC."

58. Naraian, "Disentangling the Social Threads within a Communicative Environment."

59. Roland Barthes, *The Responsibility of Forms: Critical Essays on Music, Art, and Representation*, trans. Richard Howard (Berkeley: University of California Press, 1991).

60. Martin Kevorkian, "Integrated Circuits," in *Color Monitors: The Black Face of Technology in America* (Ithaca, NY: Cornell University Press, 2006), 74–114.

61. Jonathan Sterne, *The Audible Past: Cultural Origins of Sound Reproduction* (Durham, NC: Duke University Press, 2003); Adelheid Voskuhl, "Humans, Machines, and Conversations: An Ethnographic Study of the Making of Automatic Speech Recognition Technologies," *Social Studies of Science* 34, no. 3 (2004): 393–421.

62. Pullin, "Expression Meets Information."

63. Amanda Hynan, Janice Murray, and Juliet Goldbart, "'Happy and Excited': Perceptions of Using Digital Technology and Social Media by Young People Who Use Augmentative and Alternative Communication," *Child Language Teaching and Therapy* 30, no. 2 (2014): 175–186; Amanda Hynan, Juliet Goldbart, and Janice Murray, "A Grounded Theory of Internet and Social Media Use by Young People Who Use Augmentative and Alternative Communication (AAC)," *Disability and Rehabilitation* 37, no. 17 (2015): 1559–1575; Jessica Caron and Janice Light, "'Social Media Has Opened a World of 'Open Communication': Experiences of Adults with Cerebral Palsy Who Use Augmentative and Alternative Communication and Social Media," *Augmentative and Alternative Communication* 32, no. 1 (2016): 25–40; Bronwyn Hemsley et al., "'We Definitely Need an Audience': Experiences of Twitter, Twitter Networks, and Tweet Content in Adults with Severe Communication Disabilities

Who Use Augmentative and Alternative Communication (AAC)," *Disability and Rehabilitation* 37, no. 17 (2015): 1531–1542.

64. Lacey, *Listening Publics*; Ginsburg and Rapp, "Making Disability Count"; Goggin, "Disability and the Ethics of Listening." For a discussion of the distinctions between listening and hearing in relation to media, see Lacey, *Listening Publics*, 17.

Chapter 3

1. C. Wright Mills, *The Sociological Imagination* (New York: Oxford University Press, 1959).

2. Dalton Conley, *You May Ask Yourself: An Introduction to Thinking Like a Sociologist* (New York: W. W. Norton, 2008), 4.

3. Pablo J. Boczkowski and Leah A. Lievrouw, "Bridging STS and Communication Studies: Scholarship on Media and Information Technologies," in *The Handbook of Science and Technology Studies*, ed. Edward J. Hackett et al., 3rd ed. (Cambridge, MA: MIT Press, 2008), 949–977; Paul du Gay et al., *Doing Cultural Studies: The Story of the Sony Walkman* (Thousand Oaks, CA: Sage, 1997); Tarleton Gillespie, Pablo J. Boczkowski, and Kirsten A. Foot, eds., *Media Technologies: Essays on Communication, Materiality, and Society* (Cambridge, MA: MIT Press, 2014).

4. Judy Wajcman and Paul K. Jones, "Border Communication: Media Sociology and STS," *Media, Culture, and Society* 34, no. 6 (2012): 673–690.

5. Manuel Castells et al., *Mobile Communication and Society: A Global Perspective* (Cambridge, MA: MIT Press, 2007).

6. Manuel Castells, *The Internet Galaxy: Reflections on the Internet, Business, and Society* (Oxford: Oxford University Press, 2001); Larissa Hjorth, "Odours of Mobility: Mobile Phones and Japanese Cute Culture in the Asia-Pacific," *Journal of Intercultural Studies* 26, no. 4 (2005): 39–55; James E. Katz and Satomi Sugiyama, "Mobile Phones as Fashion Statements: Evidence from Student Surveys in the US and Japan," *New Media and Society* 8, no. 2 (2006): 321–337.

7. Sherry Turkle, ed., *The Inner History of Devices* (Cambridge, MA: MIT Press, 2008).

8. Sherry Turkle, "Always-On/Always-On-You: The Tethered Self," in *Handbook of Mobile Communications and Social Change*, ed. James E. Katz (Cambridge, MA: MIT Press, 2008), 121–138.

9. Erving Goffman, *The Presentation of Self in Everyday Life* (New York: Anchor Books, 1959).

10. Mihaly Csikszentmihalyi and Eugene Rochberg-Halton, *The Meaning of Things: Domestic Symbols and the Self* (Cambridge: Cambridge University Press, 1981); David Morley, *Home Territories: Media, Mobility, and Identity* (London: Routledge, 2000).

11. Faye Ginsburg, Lila Abu-Lughod, and Brian Larkin, eds., *Media Worlds: Anthropology on New Terrain* (Berkeley: University of California Press, 2002); Heather A. Horst and Daniel Miller, *The Cell Phone: An Anthropology of Communication* (New York: Berg, 2006); Mizuko Ito, Daisuke Okabe, and Misa Matsuda, eds., *Personal, Portable, Pedestrian: Mobile Phones in Japanese Life* (Cambridge, MA: MIT Press, 2005).

12. Hjorth, "Odours of Mobility"; Katz and Sugiyama, "Mobile Phones as Fashion Statements."

13. Matthew W. Wilson, "Continuous Connectivity, Handheld Computers, and Mobile Spatial Knowledge," *Environment and Planning D: Society and Space* 32, no. 3 (2014): 535–555.

14. Michelle Rodino-Colocino, "Selling Women on PDAs from 'Simply Palm' to 'Audrey': How Moore's Law Met Parkinson's Law in the Kitchen," *Critical Studies in Media Communication* 23, no. 5 (2006): 385.

15. David Chapple, "The Evolution of Augmentative Communication and the Importance of Alternate Access," *Perspectives on Augmentative and Alternative Communication* 20, no. 1 (2011): 34–37.

16. Jean Burgess, "The iPhone Moment, the Apple Brand, and the Creative Consumer: From 'Hackability and Usability' to Cultural Generativity," in *Studying Mobile Media: Cultural Technologies, Mobile Communication, and the iPhone*, ed. Larissa Hjorth, Jean Burgess, and Ingrid Richardson (New York: Routledge, 2012), 28–42.

17. Graham Pullin, *Design Meets Disability* (Cambridge, MA: MIT Press, 2011).

18. Jessica Gosnell, John Costello, and Howard Shane, "Using a Clinical Approach to Answer 'What Communication Apps Should We Use?'" *Perspectives on Augmentatives and Alternative Communication* 20 (2011): 87–96.

19. Douglas Biklen and Jamie Burke, "Presuming Competence," *Equity and Excellence in Education* 39, no. 2 (2006): 166–175.

20. Ruth S. Cowan, *More Work for Mother: The Ironies of Household Technology from the Open Hearth to the Microwave* (New York: Basic Books, 1983); Arlie R. Hochschild, *The Managed Heart: Commercialization of Human Feeling* (Berkeley: University of California Press, 1983); Lana F. Rakow and Vija Navarro, "Remote Mothering and the Parallel Shift: Women Meet the Cellular Phone," *Critical Studies in Mass Communication* 10, no 2. (1993): 144–157; Marsha F. Cassidy, "Cyberspace Meets Domestic Space: Personal Computers, Women's Work, and the Gendered Territories of the Family Home," *Critical Studies in Media Communication* 18, no. 1 (2001): 44–65.

21. Charles C. Ragin and Howard S. Becker, eds., *What Is a Case? Exploring the Foundations of Social Inquiry* (Cambridge: Cambridge University Press, 1992).

22. Advanced Multimedia Devices, Inc., "iPad Cases for Special Needs—AMDi's iAdapter™," http://www.amdi.net/products/iadapter/iadapter-cases/iadapter-4.

23. Bruno Latour, "Technology Is Society Made Durable," in *A Sociology of Monsters: Essays on Power, Technology, and Domination*, ed. John Law (London: Routledge, 1991): 103–131.

24. Leah A. Lievrouw, "Materiality and Media in Communication and Technology Studies: An Unfinished Project," in *Media Technologies: Essays on Communication, Materiality, and Society*, ed. Tarleton Gillespie, Pablo J. Boczkowski, and Kirsten A. Foot (Cambridge, MA: MIT Press, 2014), 21–52.

25. Ingunn Moser, "Sociotechnical Practices and Difference: On the Interferences between Disability, Gender, and Class," *Science, Technology, and Human Values* 31, no. 5 (2006): 537–564.

26. Giles Slade, *Made to Break: Technology and Obsolescence in America* (Cambridge, MA: Harvard University Press, 2006).

27. Emily B. Hager, "iPad Opens World to a Disabled Boy," *New York Times*, October 31, 2010, B7.

28. Lynn Spigel, "Media Homes: Then and Now," *International Journal of Cultural Studies* 4, no. 4 (2001): 408.

29. Leopoldina Fortunati, James E. Katz, and Raimonda Riccini, eds., *Mediating the Human Body: Technology, Communication, and Fashion* (Mahwah, NJ: Lawrence Erlbaum, 2003); Marshall McLuhan, Understanding Media: The Extensions of Man, reprint ed. (Cambridge, MA: MIT Press, 1994).

30. Tom Boellstorff, *Coming of Age in Second Life: An Anthropologist Explores the Virtually Human* (Princeton, NJ: Princeton University Press, 2008); Yasmin B. Kafai, Melissa S. Cook, and Deborah A. Fields, "'Blacks Deserve Bodies Too!' Design and Discussion about Diversity and Race in a Tween Virtual World," *Games and Culture* 5, no. 1 (2010): 43–63; Lisa Nakamura, *Cybertypes: Race, Ethnicity, and Identity on the Internet* (New York: Routledge, 2002).

31. Nicola Cairns, Kevin Murray, Jonathan Corney, and Angus McFadyen, "Satisfaction with Cosmesis and Priorities for Cosmesis Design Reported by Lower Limb Amputees in the United Kingdom: Instrument Development and Results," *Prosthesis and Orthotics International* 38, no. 6 (2014): 467–473.

Chapter 4

1. Anna A. Allen and Howard C. Shane, "Autism Spectrum Disorders in the Era of Mobile Technologies: Impact on Caregivers," *Developmental Neurorehabilitation* 17, no. 2 (2014): 110–114.

2. Megan L. E. Clark, David W. Austin, and Melinda J. Craike, "Professional and Parental Attitudes Toward iPad Application Use in Autism Spectrum Disorder," *Focus on Autism and Other Developmental Disabilities* 30, no. 3 (2015): 174–181.

3. Henry Jenkins, "Introduction: 'Worship at the Altar of Convergence': A New Paradigm for Understanding Media Change," in *Convergence Culture: Where Old and New Media Collide* (New York: NYU Press, 2006), 1–24.

4. Mizuko Ito et al., *Hanging Out, Messing Around, and Geeking Out: Kids Living and Learning with New Media* (Cambridge, MA: MIT Press, 2009); Sonia Livingstone, Young People and New Media: Childhood and the Changing Media Environment (London: Sage, 2002); Ellen Seiter, *The Internet Playground: Children's Access, Entertainment, and Mis-Education* (New York: Peter Lang, 2005).

5. Victoria Rideout, *Learning at Home: Families' Educational Media Use in America* (New York: Joan Ganz Cooney Center at Sesame Workshop, 2014).

6. One family that owned an additional iPad did not report its yearly household income.

7. Rideout, *Learning at Home*.

8. The AAC specialist's sharp distinction between games and communication is likely a point of contention among members of the International Communication Association's Game Studies Division.

9. Geoffrey C. Bowker and Susan Leigh Star, *Sorting Things Out: Classification and Its Consequences* (Cambridge, MA: MIT Press, 1999).

10. Lynn Schofield Clark, *The Parent App: Understanding Families in the Digital Age* (New York: Oxford University Press, 2013); Larissa Hjorth and Sun Sun Lim, "Mobile Intimacy in an Age of Affective Mobile Media," *Feminist Media Studies* 12, no. 4 (2012): 477–484.

11. Apple, Inc., "Use Guided Access with iPhone, iPad, and iPod Touch," last modified June 16, 2015, http://support.apple.com/kb/ht5509.

12. Katherine Seelman, "Assistive Technology Policy: A Road to Independence for Individuals with Disabilities," *Journal of Social Issues* 29, no. 2 (1993): 115–136; Patricia Thornton, "Communications Technology—Empowerment or Disempowerment?" *Disability, Handicap, and Society* 8, no. 4 (1993): 339–349. It is beyond the scope of this chapter, but it should be noted that in contemporary Western culture, locks are frequently evoked as a metaphor for the relationship between the nonspeaking individual and their own body. For example, "locked-in syndrome" is a condition in which individuals cannot move or communicate through speech even though they have no loss of cognitive function. Society and the surrounding environment also serve as locks on the freedom and human rights of individuals with disabilities, yet are rarely picked or picked apart for further scrutiny.

13. Following the release of iOS 6, Apple included the Guided Access feature in successive updates to its mobile operating systems.

14. Apple Inc., "Making a Difference. One App at a Time." iOS 7 promotional video, posted June 2013, https://www.youtube.com/watch?v=3k8hUtOmcaE.

15. Jordan Golson, "iOS 6: Guided Access, Also for Kids and Kiosks," *MacRumors*, September 19, 2012, http://www.macrumors.com/2012/09/19/ios-6-feature-guided-access-also-for-kids-and-kiosks/.

16. Apple Inc., "Making a Difference."

17. Sonia Livingstone and Moira Bovill, *Families and the Internet: An Observational Study of Children and Young People's Internet Use* (London: Media@LSE, 2001).

18. Apple, Inc., "Use a Passcode with Your iPhone, iPad, or iPod Touch," last modified September 16, 2015, https://support.apple.com/en-us/HT204060.

19. It bears mentioning that the attack occurred at a regional center contracted with California's Department of Disability—a characterization that Rossmore also shares.

20. Cecilia Kang and Eric Lichtblau, "F.B.I. Error Locked San Bernardino Attacker's iPhone," *New York Times*, March 1, 2016, http://www.nytimes.com/2016/03/02/technology/apple-and-fbi-face-off-before-house-judiciary-committee.html.

21. Laurence Wilson, Joel Kaiser, and Crystal Simpson, "Final Decision Memorandum," Centers for Medicare and Medicaid Services, July 29, 2015.

22. Janice Light and David McNaughton, "Communicative Competence for Individuals Who Require Augmentative and Alternative Communication: A New Definition for a New Era of Communication?" *Augmentative and Alternative Communication* 30, no. 1 (2014): 8.

23. Meryl Alper and Beth Haller, "Social Media Use and Mediated Sociality among Individuals with Communication Disabilities in the Digital Age," in *Disability and Social Media: Global Perspectives*, ed. Katie Ellis and Mike Kent (Surrey, UK: Ashgate, forthcoming).

24. Thornton, "Communications Technology."

25. Mizuko Ito et al., *Connected Learning: An Agenda for Research and Design* (Irvine, CA: Digital Media and Learning Research Hub, 2013).

26. Pierre Bourdieu, *Outline of a Theory of Practice*, trans. Richard Nice (Cambridge: Cambridge University Press, 1972); Pierre Bourdieu, "The Forms of Capital," in *Handbook of Theory and Research for the Sociology of Education*, ed. John G. Richardson (Westport, CT: Greenwood Press, 1986), 241–258.

27. Bourdieu, *Outline of a Theory of Practice.*

28. Annette Lareau, *Home Advantage: Social Class and Parental Intervention in Elementary Education* (Oxford: Rowman and Littlefield, 2000); Jung-Sook Lee and Natasha K. Bowen, "Parent Involvement, Cultural Capital, and the Achievement Gap among Elementary School Children," *American Educational Research Journal* 43, no. 2 (2006): 193–218.

Chapter 5

1. The terms "mental retardation" and "mentally retarded" no longer appear in US health, education, and labor policy, per Rosa's Law, which President Obama signed into federal law on October 5, 2010. Those terms, emotionally charged and potentially offensive to readers, do not appear in this book outside direct quotes.

2. Jason Mittell, "The Cultural Power of an Anti-Television Metaphor: Questioning the 'Plug-in Drug' and a TV-Free America," *Television and New Media* 1, no. 2 (2000): 215–238.

3. Simi Linton, *Claiming Disability: Knowledge and Identity* (New York: NYU Press, 1998); Tobin Siebers, *Disability Theory* (Ann Arbor: University of Michigan Press, 2008).

4. Roger Desmond, "Media Literacy in the Home: Acquisition versus Deficit Models," in *Media Literacy in the Information Age: Current Perspectives*, ed. Robert Kubey (New Brunswick, NJ: Transaction Publishers, 1997), 323–343; Muriel Robinson and Bernardo Turnbull, "Verónica: An Asset Model of Becoming Literate," in *Popular Culture, New Media, and Digital Literacy in Early Childhood*, ed. Jackie Marsh (New York: Routledge Falmer, 2005), 51–72; Kathleen Tyner, *Literacy in a Digital World: Teaching and Learning in the Age of Information* (Mahwah, NJ: Lawrence Erlbaum, 1998).

5. Richard Valencia, *Dismantling Contemporary Deficit Thinking: Educational Thought and Practice* (London: Routledge, 2010).

6. Linton, *Claiming Disability*.

7. Siebers, *Disability Theory*.

8. Kristen Gillespie-Lynch et al., "Intersections between the Autism Spectrum and the Internet: Perceived Benefits and Preferred Functions of Computer-Mediated Communication," *Intellectual and Developmental Disabilities* 52, no. 6 (2014): 456–469; Scott M. Robertson, "Neurodiversity, Quality of Life, and Autistic Adults: Shifting Research and Professional Focuses onto Real-Life Challenges," *Disability Studies Quarterly* 30, no. 1 (2010), http://dsq-sds.org/article/view/1069/1234.

9. Simon Baron-Cohen, Alan M. Leslie, and Uta Frith, "Does the Autistic Child Have a 'Theory of Mind'?" *Cognition* 21 (1985): 37–46.

10. Simon Baron-Cohen, *Mindblindness: An Essay on Autism and Theory of Mind* (Cambridge, MA: MIT Press, 1997).

11. Douglas Biklen, *Autism and the Myth of the Person Alone* (New York: NYU Press, 2005).

12. Janette Dinishak and Nameera Akhtar, "A Critical Examination of Mindblindness as a Metaphor for Autism," *Child Development Perspectives* 7, no. 2 (2013): 110–114; David Smukler, "Unauthorized Minds: How 'Theory of Mind' Theory Misrepresents Autism," *Mental Retardation* 43, no. 1 (2005): 11–24.

13. John Duffy and Rebecca Dorner, "The Pathos of 'Mindblindness': Autism, Science, and Sadness in 'Theory of Mind' Narratives," *Journal of Literary & Cultural Disability Studies* 5, no. 2 (2011): 214.

14. Biklen, *Autism and the Myth of the Person Alone.*

15. Faye Ginsburg, "Disability in the Digital Age," in *Digital Anthropology*, ed. Heather A. Horst and Daniel Miller (London: Berg, 2012), 101–126; Beth Haller, *Representing Disability in an Ableist World: Essays on Mass Media* (Louisville, KY: Advocado Press, 2010).

16. Naoki Higashida, *The Reason I Jump: The Inner Voice of a Thirteen-Year-Old Boy with Autism* (New York: Random House, 2013); Ido Kedar, *Ido in Autismland: Climbing out of Autism's Silent Prison* (self-published, 2012); Arthur Fleischmann and Carly Fleischmann, *Carly's Voice: Breaking through Autism* (New York: Touchstone, 2012).

17. Desmond, "Media Literacy in the Home"; Robinson and Turnbull, "Verónica"; Tyner, *Literacy in a Digital World.*

18. Muriel Robinson and Margaret Mackey, "Film and Television," in *Handbook of Early Childhood Literacy*, ed. Joanne Larson and Jackie Marsh, 2nd ed. (Thousand Oaks, CA: Sage, 2013), 223–250; Ellen Seiter, "Power Rangers at Preschool: Negotiating Media in Child Care Settings," in *Kids' Media Culture*, ed. Marsha Kinder (Durham, NC: Duke University Press, 1999), 239–262.

19. Michael Z. Newman, "New Media, Young Audiences, and Discourses of Attention: From Sesame Street to 'Snack Culture,'" *Media, Culture, and Society* 32, no. 4 (2010): 581–596.

20. Ellen Seiter, *Sold Separately: Children and Parents in Consumer Culture* (New Brunswick, NJ: Rutgers University Press, 1995); Dennis H. Wrong, "The Oversocialized Conception of Man in Modern Sociology," *American Sociological Review* 26, no. 2 (1961); 183–193.

21. Walter Benjamin, "The Work of Art in the Age of Mechanical Reproduction," in *Illuminations: Essays and Reflections*, ed. Hannah Arendt, trans. Harry Zohn (New York: Schocken Books, 1968), 217–251; Theodor W. Adorno and Max Horkheimer, *The Dialectic of Enlightenment* (New York: Herder and Herder, 1972).

22. John Fiske, "Cultural Studies and the Culture of Everyday Life." In *Cultural Studies*, ed. Lawrence Grossberg, Cary Nelson, and Paula Treichler (New York: Routledge, 1992), 154–173; Janice Radway, *Reading the Romance: Women, Patriarchy, and Popular* Literature (Chapel Hill: University of North Carolina Press, 1984).

23. Arjun Appadurai, *Modernity at Large: Cultural Dimensions of Globalization* (Minneapolis: University of Minnesota Press, 1996); Michel de Certeau, *The Practice of Everyday Life*, trans. Steven F. Rendall (Berkeley: University of California Press, 1984); Faye Ginsburg, Lila Abu-Lughod, and Brian Larkin, eds., *Media Worlds: Anthropology on New Terrain* (Berkeley: University of California Press, 2002).

24. David Buckingham, *The Material Child: Growing Up in Consumer Culture* (Cambridge, UK: Polity Press, 2011).

25. Amit Pitaru, "E Is for Everyone: The Case for Inclusive Game Design," in *The Ecology of Games: Connecting Youth, Games, and Learning*, ed. Katie Salen (Cambridge, MA: MIT Press, 2008), 67–88.

26. Lynn Schofield Clark, *The Parent App: Understanding Families in the Digital Age* (New York: Oxford University Press, 2013); Seiter, *Sold Separately*.

27. Ron Suskind, *Life, Animated: A Story of Sidekicks, Heroes, and Autism* (Glendale, CA: Kingswell, 2014).

28. Cheryl Mattingly, "Becoming Buzz Lightyear and Other Clinical Tales: Indigenizing Disney in a World of Disability," *Folk* 45 (2003): 9–32; Cheryl Mattingly, "Pocahontas Goes to the Clinic: Popular Culture as Lingua Franca in a Cultural Borderland," *American Anthropologist* 108, no. 3 (2006): 494–501.

29. Colin Lankshear and Michele Knobel, *New Literacies: Everyday Practices and Classroom Learning*, 3rd ed. (New York: Open University Press, 2011).

30. Robinson and Turnbull, "Verónica," 68.

31. David Barton, *Literacy: An Introduction to the Ecology of Written Language* (Oxford, UK: Blackwell Publishers, 1994).

32. Adam Kendon, *Gesture: Visible Action as Utterance* (Cambridge: Cambridge University Press, 2004); Lev S. Vygotsky, *Mind in Society: The Development of Higher Psychological Processes* (Cambridge, MA: Harvard University Press, 1978).

33. Brenda Nally, Bob Houlton, and Sue Ralph, "Researches in Brief: The Management of Television and Video by Parents of Children with Autism," *Autism* 4, no. 3 (2000): 331–338; Howard C. Shane and Patti D. Albert, "Electronic Screen Media for Persons with Autism Spectrum Disorders: Results of a Survey," *Journal of Autism and Developmental Disorders* 38, no. 8 (2008): 1499–1508.

34. Marilyn Jager Adams, *Beginning to Read: Thinking and Learning about Print* (Cambridge, MA: MIT Press, 1990).

35. Miriam Liss et al., "Sensory and Attention Abnormalities in Autistic Spectrum Disorders," *Autism* 10, no. 2 (2006): 155–172; Suskind, *Life, Animated.*

36. Liss et al., "Sensory and Attention Abnormalities in Autistic Spectrum Disorders."

37. James W. Carey, "A Cultural Approach to Communication," in *Communication as Culture: Essays on Media and Society* (New York: Routledge, 1989), 13–36; Roger Silverstone, Eric Hirsch, and David Morley, "Information and Communication Technologies and the Moral Economy of the Household," in *Consuming Technologies: Media and Information in Domestic Spaces*, ed. Roger Silverstone and Eric Hirsch (London: Routledge, 1992), 15–31.

38. Meryl Alper and Beth Haller, "Social Media Use and Mediated Sociality among Individuals with Communication Disabilities in the Digital Age," in *Disability and Social Media: Global Perspectives*, ed. Katie Ellis and Mike Kent (Surrey, UK: Ashgate, forthcoming).

39. Barton, *Literacy*; Gunther Kress, *Before Writing: Rethinking the Paths to Literacy* (New York: Routledge, 1997); Barbara Rogoff, *The Cultural Nature of Human Development* (New York: Oxford University Press, 2003); Vygotsky, *Mind in Society.*

40. Jackie Marsh, ed., *Popular Culture, New Media, and Digital Literacy in Early Childhood* (New York: Routledge Falmer, 2005).

41. Kylie A. Peppler and Mark Warschauer, "Uncovering Literacies, Disrupting Stereotypes: Examining the (Dis)abilities of a Child Learning to Computer Program and Read," *International Journal of Learning and Media* 3, no. 3 (2012): 15–41.

42. Mizuko Ito, "Mobilizing the Imagination in Everyday Play: The Case of Japanese Media Mixes," in *The International Handbook of Children, Media, and Culture*, ed. Kirsten Drotner and Sonia Livingstone (London: Sage, 2008), 397–412; Becky Herr-Stephenson and Meryl Alper (with Erin Reilly and Henry Jenkins), *T Is for Transmedia: Learning through Transmedia Play* (Los Angeles: USC Annenberg Innovation Lab, 2013).

43. Martin Pistorius, *Ghost Boy: The Miraculous Escape of a Misdiagnosed Boy Trapped Inside His Own Body* (Nashville, TN: Nelson Books, 2013), ix.

44. Sherry Turkle, *Reclaiming Conversation: The Power of Talk in a Digital Age* (New York: Penguin, 2015).

45. Quoted in ibid., 37.

Chapter 6

1. Tom Streeter, "Steve Jobs, Romantic Individualism, and the Desire for Good Capitalism," *International Journal of Communication* 9 (2015): feature 3106–3124.

2. Jennifer Valentino-DeVries, "Using the iPad to Connect: Parents, Therapists Use Apple Tablet to Communicate with Special Needs Kids," *Wall Street Journal*, October 13, 2010, http://www.wsj.com/news/articles/SB10001424052748703440004575547971877769154, para. 7.

3. Ryan Budish, "What My Hearing Aid Taught Me about the Future of Wearables," *Atlantic* (blog), February 5, 2015, http://www.theatlantic.com/technology/archive/2015/02/what-my-hearing-aid-taught-me-about-the-future-of-wearables/385145/, para. 3.

4. Laurence Wilson, Joel Kaiser, and Crystal Simpson, "Final Decision Memorandum," Centers for Medicare and Medicaid Services, July 29, 2015.

5. Valentino-DeVries, "Using the iPad to Connect."

6. Anna A. Allen and Howard C. Shane, "Autism Spectrum Disorders in the Era of Mobile Technologies: Impact on Caregivers," *Developmental Neurorehabilitation* 17, no. 2 (2014): 110–114.

7. Stewart M. Hoover, Lynn Schofield Clark, and Diane Alters, *Media, Home, and Family* (New York: Routledge, 2004).

8. Elaine Lally, *At Home with Computers* (New York: Berg, 2002).

9. Henry Jenkins, "The Guiding Spirit and the Powers That Be: A Response to Suzanne Scott," in *The Participatory Cultures Handbook*, ed. Aaron Delwiche and Jennifer Jacobs Henderson (New York: Routledge, 2013), 54.

10. Faye Ginsburg, "Disability in the Digital Age," in *Digital Anthropology*, ed. Heather A. Horst and Daniel Miller (London: Berg, 2012), 101–126.

11. Rayna Rapp and Faye Ginsburg, "Reverberations: Disability and the New Kinship Imaginary," *Anthropological Quarterly* 84, no. 2 (2011): 381.

12. Elizabeth R. Lorah et al., "A Systematic Review of Tablet Computers and Portable Media Players as Speech Generating Devices for Individuals with Autism Spectrum Disorder," *Journal of Autism and Developmental Disorders* 45 (2015): 3792–3804.

13. Faye Ginsburg, Lila Abu-Lughod, and Brian Larkin, eds., *Media Worlds: Anthropology on New Terrain* (Berkeley: University of California Press, 2002).

14. Michel de Certeau, *The Practice of Everyday Life*, trans. Steven F. Rendall (Berkeley: University of California Press, 1984).

15. Arjun Appadurai, *Modernity at Large: Cultural Dimensions of Globalization* (Minneapolis: University of Minnesota Press, 1996).

16. Rapp and Ginsburg, "Reverberations."

17. Rayna Rapp and Faye Ginsburg, "Enabling Disability: Rewriting Kinship, Reimagining Citizenship," *Public Culture* 13, no. 3 (2001): 551.

18. Jennifer Cole et al., "GimpGirl Grows Up: Women with Disabilities Rethinking, Redefining, and Reclaiming Community," *New Media and Society* 13, no. 7 (2011): 1161–1179; Estelle Thoreau, "*Ouch!* An Examination of the Self-Representation of Disabled People on the Internet," *Journal of Computer-Mediated Communication* 11, no. 2 (2006): 442–468.

19. Colin Ong-Dean, "Reconsidering the Social Location of the Medical Model: An Examination of Disability in Parenting Literature," *Journal of Medical Humanities* 26, no. 2–3 (2005): 141–158.

20. Bess Williamson, "Electric Moms and Quad Drivers: People with Disabilities Buying, Making, and Using Technology in Post-War America," *American Studies* 52, no. 1 (2012): 5–29.

21. Jennifer S. Reinke and Catherine A. Solheim, "Online Social Support Experiences of Mothers of Children with Autism Spectrum Disorder," *Journal of Child and Family Studies* 24, no. 8 (2015): 2364–2373; Tawfiq Ammari, Merrie R. Morris, and Sarita Y. Schoenebeck, "Accessing Social Support and Overcoming Judgment on Social Media among Parents of Children with Special Needs," in *Proceedings of the International AAAI Conference on Weblogs and Social Media* (Palo Alto, CA: Association for the Advancement of Artificial Intelligence, 2014), 22–31; Tawfiq Ammari and Sarita Y. Schoenebeck, "Networked Empowerment on Facebook Groups for Parents of Children with Special Needs," in *Proceedings of the Annual Conference on Human Factors in Computing Systems (CHI)* (New York: ACM Press, 2015), 2805–2814.

22. Ammari and Schoenebeck, "Networked Empowerment."

23. Laura D. Zeman, Jayme Swanke, and Judy Doktor, "Strengths Classification of Social Relationships among Cybermothers Raising Children with Autism Spectrum Disorders," *School Community Journal* 21, no. 1 (2011): 37.

24. Chloe J. Jordan, "Evolution of Autism Support and Understanding via the World Wide Web," *Intellectual and Developmental Disabilities* 48, no. 3 (2009): 220–227.

25. Reinke and Solheim, "Online Social Support Experiences."

26. Ammari and Schoenebeck, "Networked Empowerment."

27. Clare Blackburn and Janet Read, "Using the Internet? The Experiences of Parents of Disabled Children," *Child: Care, Health, and Development* 31, no. 5 (2005): 507–515.

28. Lars Plantin and Kristian Daneback, "Parenthood, Information, and Support on the Internet: A Literature Review of Research on Parents and Professionals Online," *BMC Family Practice* 10, no. 34 (2009), http://www.biomedcentral.com/1471-2296/10/34; Maeve Duggan et al., *Parents and Social Media* (Washington, DC: Pew Research Center, 2015), http://www.pewinternet.org/files/2015/07/Parents-and-Social-Media-FIN-DRAFT-071515.pdf.

29. Ammari, Morris, and Schoenebeck, "Accessing Social Support."

30. Ammari and Schoenebeck, "Networked Empowerment."

31. Ammari, Morris, and Schoenebeck, "Accessing Social Support."

32. Sharon Hays, *The Cultural Contradictions of Motherhood* (New Haven, CT: Yale University Press, 1996).

33. Ong-Dean, "Reconsidering the Social Location of the Medical Model"; Colin Ong-Dean, *Distinguishing Disability: Parents, Privilege, and Special Education* (Chicago: University of Chicago Press, 2009).

34. Amy C. Sousa, "From Refrigerator Mothers to Warrior-Heroes: The Cultural Identity Transformation of Mothers Raising Children with Intellectual Disabilities," *Symbolic Interaction* 34, no. 2 (2011): 220.

35. Ong-Dean, "Reconsidering the Social Location of the Medical Model," 146.

36. Ammari, Morris, and Schoenebeck, "Accessing Social Support."

37. Virginia H. Mackintosh, Barbara J. Myers, and Robin P. Goin-Kochel, "Sources of Information and Support Used by Parents of Children with Autism Spectrum Disorders," *Journal on Developmental Disabilities* 12, no. 4 (2005): 1–51.

38. Although occurring after the completion of my fieldwork, Sesame Workshop launched an autism initiative in November 2015 that included a set of videos, with one featuring a young boy named Thomas who uses Proloquo2Go on the iPad for AAC. See http://autism.sesamestreet.org/videos/thomas-story/.

39. Margrét D. Ericsdottir and Friðrik Þór Friðriksson, *A Mother's Courage: Talking Back to Autism* (New York: First Run Features, 2010).

40. In the context of the *20/20* story, "coming out" is perhaps an allusion to the term for lesbian, gay, bisexual, transgender, and queer individuals' public self-disclosure of their sexual orientation and/or gender identity. Alternatively, it may refer to the expression for when a debutante makes her formal debut to high society.

41. Allen and Shane, "Autism Spectrum Disorders in the Era of Mobile Technologies"; Deborah Fein and Yoko Kamio, "Commentary on *The Reason I Jump* by Naoki Higashida," *Journal of Developmental and Behavioral Pediatrics* 35, no. 8 (2014): 539–542.

42. Ultimately, Ebert chose to speak in everyday conversation using a premade synthetic voice (Apple's "Alex") instead of the one made custom for him, which he reserved for recording movie reviews. See Roger Ebert, "Remaking My Voice," filmed March 2011. TED video, 19:29, http://www.ted.com/talks/roger_ebert_remaking_my_voice?language=en.

43. Jan Chapman and Jane Campion, *The Piano* (New York: Miramax Films, 1993).

44. Wikipedia, "Stephen Hawking in Popular Culture," last modified January 17, 2016, http://en.wikipedia.org/wiki/Stephen_Hawking_in_popular_culture.

45. James W. Carey, "A Cultural Approach to Communication," in *Communication as Culture: Essays on Media and Society* (New York: Routledge, 1989), 28.

46. Kathleen Kennedy, Jon Kilik, and Julian Schnabel, *The Diving Bell and the Butterfly* (New York: Miramax Films, 2007).

47. Hélène Mialet, *Hawking Incorporated: Stephen Hawking and the Anthropology of the Knowing Subject* (Chicago: University of Chicago Press, 2012).

48. Paul Jobs and Clara Jobs (née Hagopian), an Armenian American woman, adopted Steve Jobs at birth.

49. Blackburn and Read, "Using the Internet?" 514.

50. Merrie R. Morris, "Social Networking Site Use by Mothers of Young Children," in *Proceedings of the Annual Conference on Computer Supported Cooperative Work (CSCW)* (New York: ACM Press, 2014), 1272–1282; Ammari, Morris, and Schoenebeck, "Accessing Social Support"; Ammari and Schoenebeck, "Networked Empowerment."

Chapter 7

1. Mark Warschauer and Tina Matuchniak, "New Technology and Digital Worlds: Analyzing Evidence of Equity in Access, Use, and Outcomes," *Review of Research in Education* 34, no. 1 (2010): 179–225.

2. Pierre Bourdieu, *Distinction: A Social Critique of the Judgment of Taste*, trans. Richard Nice (Cambridge, MA: Harvard University Press, 1984); Pierre Bourdieu, "The Forms of Capital," in *Handbook of Theory and Research for the Sociology of Education*, ed. John G. Richardson, (Westport, CT: Greenwood Press, 1986), 241–258.

3. Michèle Lamont and Annette Lareau, "Cultural Capital: Allusions, Gaps, and Glissandos in Recent Theoretical Developments," *Sociological Theory* 6, no. 2 (1988): 153–168; Annette Lareau, *Home Advantage: Social Class and Parental Intervention in Elementary Education* (Oxford: Rowman and Littlefield, 2000); Annette Lareau, *Unequal Childhoods: Class, Race, and Family Life* (Berkeley: University of California Press, 2003); Mario L. Small, *Unanticipated Gains: Origins of Network Inequality in Everyday Life* (Oxford: Oxford University Press, 2010); Colin Ong-Dean, *Distinguishing Disability: Parents, Privilege, and Special Education* (Chicago: University of Chicago Press, 2009); Audrey A. Trainor, "Reexamining the Promise of Parent Participation in Special Education: An Analysis of Cultural and Social Capital," *Anthropology and Education Quarterly* 41, no. 3 (2010): 245–263; Lynette Kvasny, "Cultural (Re)production of Digital Inequality in a U.S. Community Technology Initiative," *Information, Communication, and Society* 9, no. 2 (2006): 160–181; Jonathan Sterne, "Bourdieu,

Technique, and Technology," *Cultural Studies* 17, no. 3–4 (2003): 367–389; Ellen Seiter, "Practicing at Home: Computers, Pianos, and Cultural Capital," in *Digital Youth, Innovation, and the Unexpected*, ed. Tara McPherson (Cambridge, MA: MIT Press, 2008), 27–52.

4. Patricia H. Collins and Valerie Chepp, "Intersectionality," in *The Oxford Handbook of Gender and Politics*, ed. Georgina Waylen et al. (New York: Oxford University Press, 2013), 57–87.

5. Quoted in Patrick Sawer, "Hawking's Fear Cuts Pose Threat for Disabled Students," *Telegraph*, May 30, 2015, http://www.telegraph.co.uk/news/science/stephen-hawking/11640763/Hawkings-fear-cuts-pose-threat-for-disabled-students.html.

6. Seth M. Holmes, *Fresh Fruit, Broken Bodies: Migrant Farmworkers in the United States* (Berkeley: University of California Press, 2013).

7. Jennifer C. Nash, "Re-Thinking Intersectionality," *Feminist Review* 89 (2008): 10.

8. Wendy Hulko, "The Time- and Context-Contingent Nature of Intersectionality and Interlocking Oppressions," *Affilia: Journal of Women and Social Work* 24, no. 1 (2009): 44–55.

9. Ange-Marie Hancock, "When Multiplication Doesn't Equal Quick Addition: Examining Intersectionality as a Research Paradigm," *Perspectives on Politics* 5, no. 1 (2007): 63–79.

10. David H. Rose and Anne Meyer, *Teaching Every Student in the Digital Age: Universal Design for Learning* (Alexandria, VA: Association for Supervision and Curriculum Development, 2002).

11. National Joint Committee for the Communication Needs of Persons with Severe Disabilities, *Guidelines for Meeting the Communication Needs of Persons with Severe Disabilities* (1992), http://www.asha.org/policy/GL1992-00201/; American Speech-Language-Hearing Association (ASHA), "Service with Culturally Diverse Individuals," n.d., http://www.asha.org/NJC/Service-With-Culturally-Diverse-Individuals/.

12. Matt Ratto and Megan Boler, eds. *DIY Citizenship: Critical Making and Social Media* (Cambridge, MA: MIT Press, 2014); Fred Turner, "Burning Man at Google: A Cultural Infrastructure for New Media Production." *New Media & Society* 11, no. 1–2 (2009): 73–94.

13. Meryl Alper, "Connecting Disability with 'Connected Learning,'" *Joan Ganz Cooney Center at Sesame Workshop* (blog), February 14, 2013, http://www.joanganzcooneycenter.org/2013/02/14/connecting-disability-with-connected-learning/.

14. Ong-Dean, *Distinguishing Disability*, 161.

15. Morgan G. Ames et al., "Understanding Technology Choices and Values through Social Class," in *Proceedings of the Annual Conference on Computer Supported Cooperative Work (CSCW)* (New York: ACM Press, 2011); Sarita Yardi and Amy Bruckman, "Income, Race, and Class: Exploring Socioeconomic Differences in Family Technology Use," in *Proceedings of the Annual Conference on Human Factors in Computing Systems (CHI)* (New York: ACM Press, 2012), 3041–3050.

16. Richard Dyer, "The Light of the World," in *White: Essays on Race and Culture* (New York: Routledge, 1997), 82–144; Lorna Roth, "Looking at Shirley, the Ultimate Norm: Colour Balance, Image Technologies, and Cognitive Equity," *Canadian Journal of Communication* 34, no. 1 (2009): 111–136.

17. Karl Wiegand and Rupal Patel, "Towards More Intelligent and Personalized AAC," Paper presented at the Supporting Children with Complex Communication Needs workshop at the annual conference on Human Factors in Computing Systems (CHI) (2014).

18. Clifford Nass and Scott Brave, *Wired for Speech: How Voice Activates and Advances the Human–Computer Relationship* (Cambridge, MA: MIT Press, 2005).

19. Kate Crawford, "Following You: Disciplines of Listening in Social Media," *Continuum: Journal of Media and Cultural Studies* 23, no. 4 (2009): 525–535.

20. Michael B. Williams, "Privacy and AAC," *Alternatively Speaking* 5, no. 2 (2000): 5.

21. Ian Hutchby, *Conversation and Technology: From the Telephone to the Internet* (Cambridge, UK: Polity, 2001).

22. Anna V. Sosa, "Association of the Type of Toy Used during Play with the Quantity and Quality of Parent–Infant Communication," *JAMA Pediatrics* 170, no. 2 (2016): 132–137; J. Light, K. D. Drager, and J. G. Nemser, "Enhancing the Appeal of AAC Technologies for Young Children: Lessons from the Toy Manufacturers," *Augmentative and Alternative Communication* 20 (2004): 137–149.

23. Graham Pullin and Shannon Hennig, "17 Ways to Say Yes: Toward Nuanced Tone of Voice in AAC and Speech Technology," *Augmentative and Alternative Communication* 31, no. 2 (2015): 170–180.

24. One expansion in the expressive potential of emoji has quite aptly come from a thirteen-year-old girl named Mercer Henderson, who has developed an emoji keyboard with accompanying sound effects. See John Brownlee, "Meet the 13-Year-Old Who's Adding Sounds to Your Emoji," *Fast Co.Design* (blog), February 17, 2016, http://www.fastcodesign.com/3056690/meet-the-13-year-old-whos-adding-sounds-to-your-emoji.

25. Steve Silberman, *NeuroTribes: The Legacy of Autism and the Future of Neurodiversity* (New York: Avery, 2015), 257–258.

26. Beth Haller, Chelsea Temple Jones, Vishaya Naidoo, Art Blaser, and Lindzey Galliford, "iTechnology as Cure or iTechnology as Empowerment: What Do North American News Media Report?" *Disability Studies Quarterly* 36, no. 1 (2016), http://dsq-sds.org/article/view/3857/4208.

27. *HillaryClinton.com*, "Factsheets: Hillary Clinton's Plan to Support Children, Youth, and Adults Living with Autism and their Families," posted January 5, 2016, https://www.hillaryclinton.com/briefing/factsheets/2016/01/05/hillary-clintons-plan-to-support-children-youth-and-adults-living-with-autism-and-their-families/.

28. Ibid.

29. Elizabeth Ellcessor, "Blurred Lines: Accessibility, Disability, and Definitional Limitations," *First Monday* 20, no. 9 (2015), http://journals.uic.edu/ojs/index.php/fm/article/view/6169/4904.

30. Katie Ellis and Gerard Goggin, *Disability and the Media* (New York: Palgrave Macmillan, 2015); Beth Haller, *Representing Disability in an Ableist World: Essays on Mass Media* (Louisville, KY: Advocado Press, 2010); Haller et al., "iTechnology as Cure or iTechnology as Empowerment."

31. GLAAD, "Where We Are on TV, 2015–2016" (Los Angeles: GLAAD, 2015), http://www.glaad.org/files/GLAAD-2015-WWAT.pdf.

32. Meryl Alper and Beth Haller, "Social Media Use and Mediated Sociality among Individuals with Communication Disabilities in the Digital Age," in *Disability and Social Media: Global Perspectives*, ed. Katie Ellis and Mike Kent (Surrey, UK: Ashgate, forthcoming).

33. Mizuko Ito, *Engineering Play: A Cultural History of Children's Software* (Cambridge, MA: MIT Press, 2009).

34. Alison Kafer, *Feminist, Queer, Crip* (Bloomington: Indiana University Press 2013), 32–33.

35. Meryl Alper, *Digital Youth with Disabilities* (Cambridge, MA: MIT Press, 2014).

36. Meryl Alper, Vikki S. Katz, and Lynn Schofield Clark, "Researching Children, Intersectionality, and Diversity in the Digital Age," *Journal of Children and Media* 10, no. 1 (2016): 107–114.

37. Jennifer A. MacMullin, Yona Lunsky, and Jonathan A. Weiss, "Plugged In: Electronics Use in Youth and Young Adults with Autism Spectrum Disorder," *Autism* 20, no. 1 (2016): 45–54.

38. Benjamin Zablotsky et al., "Estimated Prevalence of Autism and Other Developmental Disabilities Following Questionnaire Changes in the 2014 National Health Interview Survey," *National Health Statistics Reports, No. 87* (Hyattsville, MD: National Center for Health Statistics, 2015).

39. Jordynn Jack, *Autism and Gender: From Refrigerator Mothers to Computer Geeks* (Champaign: University of Illinois Press, 2014).

40. Gareth Cook, "The Autism Advantage," *New York Times Magazine*, December 2, 2012, http://www.nytimes.com/2012/12/02/magazine/the-autism-advantage.html.

41. Sunny Taylor, "The Right Not to Work: Power and Disability," *Monthly Review* 55, no. 10 (2004), http://monthlyreview.org/2004/03/01/the-right-not-to-work-power-and-disability/.

42. Silberman, *NeuroTribes*.

43. Paul T. Shattuck et al., "Postsecondary Education and Employment among Youth with an Autism Spectrum Disorder," *Pediatrics* 129, no. 6 (2012): 1042–1049.

44. Kristen Gillespie-Lynch et al., "Intersections between the Autism Spectrum and the Internet: Perceived Benefits and Preferred Functions of Computer-Mediated Communication," *Intellectual and Developmental Disabilities* 52, no. 6 (2014): 456–469; Micah O. Mazurek and Christopher R. Engelhardt, "Video Game Use and Problem Behaviors in Boys with Autism Spectrum Disorders," *Research in Autism Spectrum Disorders* 7, no. 2 (2013): 316–324; Micah O. Mazurek and Christopher R. Engelhardt. "Video Game Use in Boys with Autism Spectrum Disorder, ADHD, or Typical Development," *Pediatrics* 132, no. 2 (2013): 260–266; Howard C. Shane and Patti D. Albert, "Electronic Screen Media for Persons with Autism Spectrum Disorders: Results of a Survey," *Journal of Autism and Developmental Disorders* 38, no. 8 (2008): 1499–1508; Erinn H. Finke, Benjamin Hickerson, and Eileen McLaughlin, "Parental Intention to Support Video Game Play by Children with Autism Spectrum Disorder: An Application of the Theory of Planned Behavior," *Language, Speech, and Hearing Services in Schools* 46 (2015): 154–165.

Methods

1. Ericka L. Wodka, Pamela Mathy, and Luther Kalb, "Predictors of Phrase and Fluent Speech in Children with Autism and Severe Language Delay," *Pediatrics* 131, no 4 (2013): 1128–1134.

2. Fritjof Norrelgen et al., "Children with Autism Spectrum Disorders Who Do Not Develop Phrase Speech in the Preschool Years," *Autism* 19, no. 8 (2015): 934–943.

3. Meryl Alper, *Digital Youth with Disabilities* (Cambridge, MA: MIT Press, 2014).

4. Tragically, one of California's regional centers made worldwide headlines in December 2015 when the Inland Regional Center in San Bernardino was the site of a mass shooting.

5. For additional information on the history and current state of California's regional center system, see Association of Regional Center Agencies, *On the Brink of*

Collapse: The Consequences of Underfunding California's Developmental Services System (Sacramento: Association of Regional Center Agencies, 2015), http://arcanet.org/wp-content/uploads/2015/02/on-the-brink-of-collapse.pdf.

6. Morgan G. Ames et al., "Understanding Technology Choices and Values through Social Class," in *Proceedings of the Annual Conference on Computer Supported Cooperative Work (CSCW)* (New York: ACM Press, 2011), 55–64.

7. Thomas R. Lindlof and Bryan C. Taylor, *Qualitative Communication Research Methods*, 2nd ed. (Thousand Oaks, CA: Sage, 2002), 123.

8. Steven J. Taylor and Robert Bogdan, *Introduction to Qualitative Research Methods*, 3rd ed. (New York: Wiley, 1998).

9. Kathy Charmaz, "The Grounded Theory Method: An Explication and Interpretation," in *Contemporary Field Research*, ed. Robert M. Emerson (Boston: Little Brown, 1983), 109–126; Barney G. Glaser and Anselm L. Strauss, *The Discovery of Grounded Theory: Strategies for Qualitative Research* (Chicago: Aldine Publishing Company, 1967); Anselm L. Strauss and Juliet Corbin, *Basics of Qualitative Research: Techniques and Procedures for Developing Grounded Theory*, 2nd ed. (London: Sage, 1998).

10. Jordynn Jack, *Autism and Gender: From Refrigerator Mothers to Computer Geeks* (Champaign: University of Illinois Press, 2014).

Bibliography

Adams, Marilyn Jager. *Beginning to Read: Thinking and Learning about Print.* Cambridge, MA: MIT Press, 1990.

Adams, Rachel. *Raising Henry: A Memoir of Motherhood, Disability, and Discovery*. New Haven, CT: Yale University Press, 2013.

Adorno, Theodor W., and Max Horkheimer. *The Dialectic of Enlightenment.* New York: Herder and Herder, 1972.

Advanced Multimedia Devices, Inc. "iPad Cases for Special Needs—AMDi's iAdapter™." http://www.amdi.net/products/iadapter/iadapter-cases/iadapter-4.

Allen, Anna A., and Howard C. Shane. "Autism Spectrum Disorders in the Era of Mobile Technologies: Impact on Caregivers." *Developmental Neurorehabilitation* 17, no. 2 (2014): 110–114.

Allen, James P., and Eugene Turner. "Ethnic Change and Enclaves in Los Angeles." *American Association of Geographers* (blog), March 8, 2013. http://www.aag.org/cs/news_detail?pressrelease.id=2058.

Alper, Meryl. "Connecting Disability with 'Connected Learning.'" *Joan Ganz Cooney Center at Sesame Workshop* (blog), February 14, 2013. http://www.joanganzcooneycenter.org/2013/02/14/connecting-disability-with-connected-learning/.

Alper, Meryl. *Digital Youth with Disabilities*. Cambridge, MA: MIT Press, 2014.

Alper, Meryl. When Face-to-Face Is Screen-to-Screen: Reconsidering Mobile Media as Communication Augmentations and Alternatives. In *Routledge Companion to Disability and Media*, ed. Katie Ellis, Gerard Goggin, and Beth Haller. New York: Routledge, 2017.

Alper, Meryl, and Beth Haller. Social Media Use and Mediated Sociality among Individuals with Communication Disabilities in the Digital Age. In *Disability and Social Media: Global Perspectives*, ed. Katie Ellis and Mike Kent. Surrey, UK: Ashgate, forthcoming.

Alper, Meryl, Vikki S. Katz, and Lynn Schofield Clark. "Researching Children, Intersectionality, and Diversity in the Digital Age." *Journal of Children and Media* 10, no. 1 (2016): 107–114.

American Speech-Language-Hearing Association (ASHA). *Communication Facts: Special Populations: Augmentative and Alternative Communication, 2008 Edition*. Rockville, MD: ASHA, 2008.

American Speech-Language-Hearing Association (ASHA). "Childhood Apraxia of Speech." n.d. http://www.asha.org/public/speech/disorders/ChildhoodApraxia/.

American Speech-Language-Hearing Association (ASHA). "Funding for Services." n.d. http://www.asha.org/NJC/Funding-for-Services/.

American Speech-Language-Hearing Association (ASHA). "Service with Culturally Diverse Individuals." n.d. http://www.asha.org/NJC/Service-With-Culturally-Diverse-Individuals/.

Ames, Morgan G., Janet Go, Joseph "Jofish" Kaye, and Mirjana Spasojevic. "Understanding Technology Choices and Values through Social Class." In *Proceedings of the Annual Conference on Computer Supported Cooperative Work (CSCW)*, 55–64. New York: ACM Press, 2011.

Ammari, Tawfiq, Merrie R. Morris, and Sarita Y. Schoenebeck. "Accessing Social Support and Overcoming Judgment on Social Media among Parents of Children with Special Needs." In *Proceedings of the International AAAI Conference on Weblogs and Social Media*, 22–31. Palo Alto, CA: Association for the Advancement of Artificial Intelligence, 2014.

Ammari, Tawfiq, and Sarita Y. Schoenebeck. "Networked Empowerment on Facebook Groups for Parents of Children with Special Needs." In *Proceedings of the Annual Conference on Human Factors in Computing Systems (CHI)*, 2805–2814. New York: ACM Press, 2015.

Angelo, Dianne H. AAC in the Family and Home. In *The Handbook of Augmentative and Alternative Communication*, ed. Sharon Glennen and Denise C. DeCoste, 523–541. San Diego: Singular Publishing Group, 1997.

Appadurai, Arjun. Introduction: Commodities and the Politics of Value. In *The Social Life of Things: Commodities in Cultural Perspective*, ed. Arjun Appadurai, 3–63. Cambridge, UK: Cambridge University Press, 1986.

Appadurai, Arjun. *Modernity at Large: Cultural Dimensions of Globalization*. Minneapolis: University of Minnesota Press, 1996.

Apple, Inc. "Making a Difference. One App at a Time." iOS 7 promotional video. Posted June 2013. https://www.youtube.com/watch?v=3k8hUtOmcaE.

Apple, Inc. "Use a Passcode with Your iPhone, iPad, or iPod Touch." Last modified September 16, 2015. https://support.apple.com/en-us/HT204060.

Apple, Inc. "Use Guided Access with iPhone, iPad, and iPod Touch." Last modified June 16, 2015. http://support.apple.com/kb/ht5509.

Apple, Inc. "Volume Purchase Program: Proloquo2Go." n.d. https://volume.itunes.apple.com/us/app/proloquo2go-symbol-based-aac/id308368164?mt=8&term=Proloquo2Go&ign-mpt=uo%3D4.

Apple, Michael W. *Teachers and Texts: A Political Economy of Class and Gender Relations in Education*. New York: Routledge, 1986.

Apple, Michael W., and Linda Christian-Smith, eds. *The Politics of the Textbook*. New York: Routledge, 1991.

Ashby, Christine. "Whose 'Voice' Is It Anyway?: Giving Voice and Qualitative Research Involving Individuals That Type to Communicate." *Disability Studies Quarterly* 31, no. 4 (2011). http://dsq-sds.org/article/view/1723/1771.

AssistiveWare. "AssistiveWare—The Making of the American Children's Voices Josh and Ella." YouTube video, 1:45. Posted July 25, 2012. https://www.youtube.com/watch?v=OsI17MZIq0U.

AssistiveWare. "Proloquo2Go Flyer." 2013 . http://download.assistiveware.com/proloquo2go/files/Proloquo2Go_Flyer.pdf.

AssistiveWare. "Working with Recents." 2013. http://download.assistiveware.com/proloquo2go/files/version3/Proloquo2Go_3_Tutorial-Working_with_Recents.pdf.

AssistiveWare. "World's First Bilingual Spanish–English Children's Text to Speech Voices." Posted January 28, 2014. http://www.assistiveware.com/worlds-first-bilingual-spanish-english-childrens-text-speech-voices.

AssistiveWare. "Award-Winning Communication App Now Gives a Voice to Spanish Children." Posted September 30, 2015. https://www.assistiveware.com/award-winning-communication-app-now-gives-voice-spanish-children.

AssistiveWare. *Proloquo2Go Manual Version 4.1*. Amsterdam: AssistiveWare, 2015. http://download.assistiveware.com/proloquo2go/helpfiles/4.1/en/assets/Proloquo2GoManual.pdf.

AssistiveWare. "Proloquo2Go." n.d. http://www.assistiveware.com/product/proloquo2go.

AssistiveWare. "Voices: U.S. English." n.d. http://www.assistiveware.com/product/proloquo2go/voices#en_US.

Association of Regional Center Agencies. *On the Brink of Collapse: The Consequences of Underfunding California's Developmental Services System*. Sacramento: Association of

Regional Center Agencies, 2015. http://arcanet.org/wp-content/uploads/2015/02/on-the-brink-of-collapse.pdf.

Attewell, Paul, and Juan Battle. "Home Computers and School Performance." *Information Society* 15, no. 1 (1999): 1–10.

Axel, Brian K. "Anthropology and the New Technologies of Communication." *Cultural Anthropology* 21, no. 3 (2006): 354–384.

Bailey, Rita L., Howard P. Parette Jr., Julia B. Stoner, Maureen E. Angell, and Kathleen Carroll. "Family Members' Perceptions of Augmentative and Alternative Communication Device Use." *Language, Speech, and Hearing Services in Schools* 37 (2006): 50–60.

Bakardjieva, Maria, and Richard Smith. "The Internet in Everyday Life: Computer Networking from the Standpoint of the Domestic User." *New Media and Society* 3, no. 1 (2001): 67–83.

Banet-Weiser, Sarah, and Kate Miltner. "#MasculinitySoFragile: Culture, Structure, and Networked Misogyny." *Feminist Media Studies* 16, no. 1 (2016): 171–174.

Baron-Cohen, Simon. *Mindblindness: An Essay on Autism and Theory of Mind.* Cambridge, MA: MIT Press, 1997.

Baron-Cohen, Simon, Alan M. Leslie, and Uta Frith. "Does the Autistic Child Have a 'Theory of Mind'?" *Cognition* 21 (1985): 37–46.

Barthes, Roland. *The Responsibility of Forms: Critical Essays on Music, Art, and Representation.* Trans. Richard Howard. Berkeley: University of California Press, 1991.

Barton, David. *Literacy: An Introduction to the Ecology of Written Language.* Oxford, UK: Blackwell Publishers, 1994.

Baym, Nancy. *Personal Connections in the Digital Age.* 2nd ed. Cambridge, UK: Polity Press, 2015.

Benjamin, Walter. The Work of Art in the Age of Mechanical Reproduction. In *Illuminations: Essays and Reflections*, ed. Hannah Arendt, trans. Harry Zohn, 217–251. New York: Schocken Books, 1968.

Bernheimer, Lucinda P., Ronald Gallimore, and Thomas S. Weisner. "Ecocultural Theory as a Context for the Individual Family Service Plan." *Journal of Early Intervention* 14, no. 3 (1990): 219–233.

Bernstein, Basil. *Class, Codes, and Control.* 2nd ed. Vol. 3. London: Routledge and Kegan Paul, 1977.

Bettelheim, Bruno. "Joey: A 'Mechanical Boy.'" *Scientific American* 200 (1959): 116–127.

Bettelheim, Bruno. *The Empty Fortress: Infantile Autism and the Birth of the Self.* New York: Free Press, 1967.

Beukelman, David, and Pat Mirenda. *Augmentative and Alternative Communication: Supporting Children and Adults with Complex Communication Needs.* 4th ed. Baltimore: Brookes Publishing, 2013.

Bickford, Susan. *The Dissonance of Democracy: Listening, Conflict, and Citizenship.* Ithaca, NY: Cornell University Press, 1996.

Biklen, Douglas. *Autism and the Myth of the Person Alone.* New York: NYU Press, 2005.

Biklen, Douglas, and Jamie Burke. "Presuming Competence." *Equity and Excellence in Education* 39, no. 2 (2006): 166–175.

Blackburn, Clare, and Janet Read. "Using the Internet? The Experiences of Parents of Disabled Children." *Child: Care, Health, and Development* 31, no. 5 (2005): 507–515.

Blum, Linda. *Raising Generation Rx: Mothering Kids with Invisible Disabilities in an Age of Inequality.* New York: NYU Press, 2015.

Boczkowski, Pablo J., and Leah A. Lievrouw. Bridging STS and Communication Studies: Scholarship on Media and Information Technologies. In *The Handbook of Science and Technology Studies,* ed. Edward J. Hackett, Olga Amsterdamska, Michael E. Lynch, and Judy Wajcman, 949–977. 3rd ed. Cambridge, MA: MIT Press, 2008.

Boellstorff, Tom. *Coming of Age in Second Life: An Anthropologist Explores the Virtually Human.* Princeton, NJ: Princeton University Press, 2008.

Bourdieu, Pierre. *Outline of a Theory of Practice.* Trans. Richard Nice. Cambridge, UK: Cambridge University Press, 1972.

Bourdieu, Pierre. *Distinction: A Social Critique of the Judgment of Taste.* Trans. Richard Nice. Cambridge, MA: Harvard University Press, 1984.

Bourdieu, Pierre. The Forms of Capital. In *Handbook of Theory and Research for the Sociology of Education,* ed. John G. Richardson, 241–258. Westport, CT: Greenwood Press, 1986.

Bourdieu, Pierre, and Loïc J. D. Wacquant. *An Invitation to Reflexive Sociology.* Chicago: University of Chicago Press, 1992.

Bowker, Geoffrey C., and Susan Leigh Star. *Sorting Things Out: Classification and Its Consequences.* Cambridge, MA: MIT Press, 1999.

boyd, danah. *It's Complicated: The Social Lives of Networked Teens.* New Haven, CT: Yale University Press, 2014.

Boyle, Coleen A., Sheree Boulet, Laura A. Schieve, Robin A. Cohen, Stephen J. Blumberg, Marshalyn Yeargin-Allsopp, Susanna Visser, and Michael D. Kogan. "Trends in the Prevalence of Developmental Disabilities in U.S. Children, 1997–2008." *Pediatrics* 127 (2011): 1034–1042.

Brownlee, John. "Meet the 13-Year-Old Who's Adding Sounds to Your Emoji." *Fast Co.Design* (blog), February 17, 2016. http://www.fastcodesign.com/3056690/meet-the-13-year-old-whos-adding-sounds-to-your-emoji.

Buckingham, David. *The Material Child: Growing Up in Consumer Culture*. Cambridge, UK: Polity Press, 2011.

Buckingham, David, and Margaret Scanlon. *Education, Entertainment, and Learning in the Home*. Philadelphia: Open University Press, 2003.

Budish, Ryan. "What My Hearing Aid Taught Me about the Future of Wearables." *Atlantic* (blog), February 5, 2015. http://www.theatlantic.com/technology/archive/2015/02/what-my-hearing-aid-taught-me-about-the-future-of-wearables/385145/.

Burgess, Jean. "Hearing Ordinary Voices: Cultural Studies, Vernacular Creativity, and Digital Storytelling." *Continuum: Journal of Media and Cultural Studies* 20, no. 2 (2006): 201–214.

Burgess, Jean. The iPhone Moment, the Apple Brand, and the Creative Consumer: From "Hackability and Usability" to Cultural Generativity. In *Studying Mobile Media: Cultural Technologies, Mobile Communication, and the iPhone*, ed. Larissa Hjorth, Jean Burgess, and Ingrid Richardson, 28–42. New York: Routledge, 2012.

Butler, Judith. *Gender Trouble: Feminism and the Subversion of Identity*. New York: Routledge, 1990.

Cairns, Nicola, Kevin Murray, Jonathan Corney, and Angus McFadyen. "Satisfaction with Cosmesis and Priorities for Cosmesis Design Reported by Lower Limb Amputees in the United Kingdom: Instrument Development and Results." *Prosthetics and Orthotics International* 38, no. 6 (2014): 467–473.

Carey, James W. A Cultural Approach to Communication. In *Communication as Culture: Essays on Media and Society*, 13–36. New York: Routledge, 1989.

Carmichael, Callie. "How Tablets Helped Unlock One Girl's Voice." *CNN* (blog). November 14, 2012. http://www.cnn.com/2012/11/14/tech/mobile/carly-fleischmann-mobile-autism/.

Caron, Jessica, and Janice Light. "'Social Media Has Opened a World of 'Open Communication': Experiences of Adults with Cerebral Palsy Who Use Augmentative and Alternative Communication and Social Media." *Augmentative and Alternative Communication* 32, no. 1 (2016): 25–40.

Carter, Elizabeth J., Diane L. Williams, Jessica K. Hodgins, and Jill F. Lehman. "Are Children with Autism More Responsive to Animated Characters? A Study of

Interactions with Humans and Human-Controlled Avatars." *Journal of Autism and Developmental Disorders* 44, no. 10 (2014): 2475–2485.

Cartwright, Lisa. *Moral Spectatorship: Technologies of Voice and Affect in Postwar Representations of the Child*. Durham, NC: Duke University Press, 2008.

Cassidy, Marsha F. "Cyberspace Meets Domestic Space: Personal Computers, Women's Work, and the Gendered Territories of the Family Home." *Critical Studies in Media Communication* 18, no. 1 (2001): 44–65.

Castells, Manuel. *The Internet Galaxy: Reflections on the Internet, Business, and Society*. Oxford: Oxford University Press, 2001.

Castells, Manuel, Mireia Fernández-Ardèvol, Jack Linchuan Qiu, and Araba Sey. *Mobile Communication and Society: A Global Perspective*. Cambridge, MA: MIT Press, 2007.

Chapman, Jan, and Jane Campion. *The Piano*. New York: Miramax Films, 1993.

Chapple, David. "The Evolution of Augmentative Communication and the Importance of Alternate Access." *Perspectives on Augmentative and Alternative Communication* 20, no. 1 (2011): 34–37.

Charlton, James I. *Nothing about Us without Us: Disability Oppression and Empowerment*. Berkeley: University of California Press, 2000.

Charmaz, Kathy. The Grounded Theory Method: An Explication and Interpretation. In *Contemporary Field Research*, ed. Robert M. Emerson, 109–126. Boston: Little Brown, 1983.

Clark, Lynn Schofield. "Parental Mediation Theory for the Digital Age." *Communication Theory* 21, no. 4 (2011): 323–343.

Clark, Lynn Schofield. *The Parent App: Understanding Families in the Digital Age*. New York: Oxford University Press, 2013.

Clark, Megan L. E., David W. Austin, and Melinda J. Craike. "Professional and Parental Attitudes toward iPad Application Use in Autism Spectrum Disorder." *Focus on Autism and Other Developmental Disabilities* 30, no. 3 (2015): 174–181.

Cole, Jennifer, Jason Nolan, Yukari Seko, Katherine Mancuso, and Alejandra Ospina. "GimpGirl Grows Up: Women with Disabilities Rethinking, Redefining, and Reclaiming Community." *New Media and Society* 13, no. 7 (2011): 1161–1179.

Collins, Patricia H. *Fighting Words: Black Women and the Search for Justice*. Minneapolis: University of Minneapolis Press, 1998.

Collins, Patricia H. "Reply to Commentaries. *Black Sexual Politics* Revisited." *Studies in Gender and Sexuality* 9, no. 1 (2008): 68–85.

Collins, Patricia H., and Valerie Chepp. Intersectionality. In *The Oxford Handbook of Gender and Politics*, ed. Georgina Waylen, Karen Celis, Johanna Kantola, and S. Laurel Weldon, 57–87. New York: Oxford University Press, 2013.

Conley, Dalton. *You May Ask Yourself: An Introduction to Thinking Like a Sociologist.* New York: W. W. Norton, 2008.

Cook, Gareth. "The Autism Advantage." *New York Times Magazine*, December 2, 2012. http://www.nytimes.com/2012/12/02/magazine/the-autism-advantage.html.

Costanza-Chock, Sasha. "Mic Check! Media Cultures and the Occupy Movement." *Social Movement Studies: Journal of Social, Cultural, and Political Protest* 11, no. 3–4 (2012): 1–11.

Costanza-Chock, Sasha. *Out of the Shadows, Into the Streets: Transmedia Organizing and the Immigrant Rights Movement.* Cambridge, MA: MIT Press, 2014.

Couldry, Nick. "Rethinking the Politics of Voice." *Continuum: Journal of Media and Cultural Studies* 23, no. 4 (2009): 579–582.

Couldry, Nick. *Why Voice Matters: Culture and Politics after Neoliberalism.* London: Sage, 2010.

Couldry, Nick. Voiceblind: Beyond the Paradoxes of the Neoliberal State. In *Reclaiming the Public Sphere: Communication, Power, and Social Change*, ed. Tina Askanius and Liv S. Østergaard, 15–25. New York: Palgrave Macmillan, 2014.

Cowan, Ruth S. *More Work for Mother: The Ironies of Household Technology from the Open Hearth to the Microwave.* New York: Basic Books, 1983.

Crawford, Kate. "Following You: Disciplines of Listening in Social Media." *Continuum: Journal of Media and Cultural Studies* 23, no. 4 (2009): 525–535.

Crawford, Kate. Four Ways of Listening to an iPhone: From Sound and Network Listening to Biometric Data and Geolocative Tracking. In *Studying Mobile Media: Cultural Technologies, Mobile Communication, and the iPhone*, ed. Larissa Hjorth, Jean Burgess, and Ingrid Richardson, 213–239. New York: Routledge, 2012.

Creer, Sarah, Stuart Cunningham, Phil Green, and Yamagishi Junichi. "Building Personalised Synthetic Voices for Individuals with Severe Speech Impairment." *Computer Speech and Language* 27, no. 6 (2013): 1178–1193.

Crenshaw, Kimberlé. "Demarginalizing the Intersection of Race and Sex: A Black Feminist Critique of Antidiscrimination Doctrine, Feminist Theory, and Antiracist Politics." *University of Chicago Legal Forum* 1 (1989): 139–167.

Crenshaw, Kimberlé. "Mapping the Margins: Intersectionality, Identity Politics, and Violence against Women of Color." *Stanford Law Review* 43, no. 6 (1991): 1241–1299.

Csikszentmihalyi, Mihaly, and Eugene Rochberg-Halton. *The Meaning of Things: Domestic Symbols and the Self.* Cambridge, UK: Cambridge University Press, 1981.

Cuban, Larry. *Oversold and Underused: Computers in Classrooms, 1980–2000.* Cambridge, MA: Harvard University Press, 2001.

Davis, Lennard J. *Bending Over Backwards: Disability, Dismodernism, and Other Difficult Positions.* New York: NYU Press, 2002.

Dawe, Melissa. "Desperately Seeking Simplicity: How Young Adults with Cognitive Disabilities and Their Families Adopt Assistive Technologies." In *Proceedings of the Annual Conference on Human Factors in Computing Systems (CHI)*, 1143–1152. New York: ACM Press, 2006.

de Certeau, Michel. *The Practice of Everyday Life.* Trans. Steven F. Rendall. Berkeley: University of California Press, 1984.

Derrida, Jacques. *Voice and Phenomenon: Introduction to the Problem of the Sign in Husserl's Phenomenology.* Trans. Leonard Lawlor. Evanston, IL: Northwestern University Press, 2011.

Desmond, Roger. Media Literacy in the Home: Acquisition versus Deficit Models. In *Media Literacy in the Information Age: Current Perspectives*, ed. Robert Kubey, 323–343. New Brunswick, NJ: Transaction Publishers, 1997.

Diep, Lucy, and Gregor Wolbring. "Perceptions of Brain–Machine Interface Technology among Mothers of Disabled Children." *Disability Studies Quarterly* 35, no. 4 (2015). http://dsq-sds.org/article/view/3856/4112.

Dinishak, Janette, and Nameera Akhtar. "A Critical Examination of Mindblindness as a Metaphor for Autism." *Child Development Perspectives* 7, no. 2 (2013): 110–114.

Dobransky, Kerry, and Eszter Hargittai. "The Disability Divide in Internet Access and Use." *Information, Communication, and Society* 9, no. 3 (2006): 313–334.

Dobson, Andrew. "Democracy and Nature: Speaking and Listening." *Political Studies* 58, no. 4 (2010): 752–768.

Dolar, Mladen. *A Voice and Nothing More.* Cambridge, MA: MIT Press, 2006.

Donner, Jonathan. *After Access: Inclusion, Development, and a More Mobile Internet.* Cambridge, MA: MIT Press, 2015.

Dreher, Tanja. "Listening across Difference: Media and Multiculturalism beyond the Politics of Voice." *Journal of Media and Cultural Studies* 23, no. 4 (2009): 445–458.

Dreher, Tanja. "A Partial Promise of Voice: Digital Storytelling and the Limit of Listening." *Media International Australia* 142, no. 1 (2012): 157–166.

Du Bois, W.E.B. *Black Reconstruction in America, 1860–1880.* New York: Atheneum, 1969.

Duffy, John, and Rebecca Dorner. "The Pathos of 'Mindblindness': Autism, Science, and Sadness in 'Theory of Mind' Narratives." *Journal of Literary and Cultural Disability Studies* 5, no. 2 (2011): 201–216.

du Gay, Paul, Stuart Hall, Linda Janes, Hugh Mackay, and Keith Negus. *Doing Cultural Studies: The Story of the Sony Walkman*. Thousand Oaks, CA: Sage, 1997.

Duggan, Maeve, Amanda Lenhart, Cliff Lampe, and Nicole B. Ellison. *Parents and Social Media*. Washington, DC: Pew Research Center, 2015. http://www.pewinternet.org/files/2015/07/Parents-and-Social-Media-FIN-DRAFT-071515.pdf.

Dunbar-Hester, Christina. *Low Power to the People: Pirates, Protest, and Politics in FM Radio Activism*. Cambridge, MA: MIT Press, 2014.

Dybwad, Barb. "2-Year-Old Finds iPad Easy to Use." *Mashable*, April 6, 2010. http://mashable.com/2010/04/06/2-year-old-girl-uses-ipad/.

Dyer, Richard. The Light of the World. In *White: Essays on Race and Culture*, 82–144. New York: Routledge, 1997.

Ebert, Roger. "Remaking My Voice." Filmed March 2011. TED video, 19:29. http://www.ted.com/talks/roger_ebert_remaking_my_voice?language=en.

Ellcessor, Elizabeth. "Blurred Lines: Accessibility, Disability, and Definitional Limitations." *First Monday* 20, no. 9 (2015). http://journals.uic.edu/ojs/index.php/fm/article/view/6169/4904.

Ellcessor, Elizabeth. *Restricted Access: Media, Disability, and the Politics of Participation*. New York: NYU Press, 2016.

Ellis, Katie, and Gerard Goggin. *Disability and the Media*. New York: Palgrave Macmillan, 2015.

Ellis, Katie, and Mike Kent. *Disability and New Media*. New York: Routledge, 2010.

Elman, Julie P. *Chronic Youth: Disability, Sexuality, and US Media Cultures of Rehabilitation*. New York: NYU Press, 2014.

Ericsdottir, Margrét D., and Friðrik Þór Friðriksson. *A Mother's Courage: Talking Back to Autism*. New York: First Run Features, 2010.

Everett, Anna, ed. *Learning Race and Ethnicity: Youth and Digital Media*. Cambridge, MA: MIT Press, 2008.

Fein, Deborah, and Yoko Kamio. "Commentary on *The Reason I Jump* by Naoki Higashida." *Journal of Developmental and Behavioral Pediatrics* 35, no. 8 (2014): 539–542.

Fernandes, Barbara. "iTherapy: The Revolution of Mobile Devices within the Field of Speech Therapy." *Perspectives on School-Based Issues* 12, no. 2 (2011): 35–40.

Finke, Erinn H., Benjamin Hickerson, and Eileen McLaughlin. "Parental Intention to Support Video Game Play by Children with Autism Spectrum Disorder: An Application of the Theory of Planned Behavior." *Language, Speech, and Hearing Services in Schools* 46 (2015): 154–165.

Fiske, John. Cultural Studies and the Culture of Everyday Life. In *Cultural Studies*, ed. Lawrence Grossberg, Cary Nelson and Paula Treichler, 154–173. New York: Routledge, 1992.

Flanagan, James L. "Voices of Men and Machines." *Journal of the Acoustical Society of America* 51, no. 5 (1972): 1375–1387.

Fleischmann, Arthur, and Carly Fleischmann. *Carly's Voice: Breaking through Autism*. New York: Touchstone, 2012.

Fortunati, Leopoldina. Real People, Artificial Bodies. In *Mediating the Human Body: Technology, Communication, and Fashion*, ed. Leopoldina Fortunati, James E. Katz, and Raimonda Riccini, 61–71. Mahwah, NJ: Lawrence Erlbaum, 2003.

Fortunati, Leopoldina, James E. Katz, and Raimonda Riccini, eds. *Mediating the Human Body: Technology, Communication, and Fashion*. Mahwah, NJ: Lawrence Erlbaum, 2003.

Foucault, Michel. "Technologies of the Self." In *Technologies of the Self: A Seminar with Michel Foucault*, ed. Huck Gutman, Patrick H. Hutton, and Luther H. Martin, 16–49. Amherst: University of Massachusetts Press, 1988.

Friedman, Sandra L., and Miriam A. Kalichman. "Out-of-Home Placement for Children and Adolescents with Disabilities." *Pediatrics* 134, no. 4 (2014): 836–846.

Garland-Thomson, Rosemarie. *Extraordinary Bodies: Figuring Physical Disability in American Culture and Literature*. New York: Columbia University Press, 1996.

Gillespie, Tarleton. The Relevance of Algorithms. In *Media Technologies: Essays on Communication, Materiality, and Society*, ed. Tarleton Gillespie, Pablo J. Boczkowski, and Kirsten A. Foot, 167–194. Cambridge, MA: MIT Press, 2014.

Gillespie, Tarleton, Pablo J. Boczkowski, and Kirsten A. Foot, eds. *Media Technologies: Essays on Communication, Materiality, and Society*. Cambridge, MA: MIT Press, 2014.

Gillespie-Lynch, Kristen, Steven K. Kapp, Christina Shane-Simpson, David Shane Smith, and Ted Hutman. "Intersections between the Autism Spectrum and the Internet: Perceived Benefits and Preferred Functions of Computer-Mediated Communication." *Intellectual and Developmental Disabilities* 52, no. 6 (2014): 456–469.

Ginsburg, Faye. Disability in the Digital Age. In *Digital Anthropology*, ed. Heather A. Horst and Daniel Miller, 101–126. London: Berg, 2012.

Ginsburg, Faye, Lila Abu-Lughod, and Brian Larkin, eds. *Media Worlds: Anthropology on New Terrain*. Berkeley: University of California Press, 2002.

Ginsburg, Faye, and Rayna Rapp. "Making Disability Count: Demography, Futurity, and the Making of Disability Publics." *Somatosphere*, May 11, 2015. http://somatosphere.net/2015/05/making-disability-count-demography-futurity-and-the-making-of-disability-publics.html.

Gitelman, Lisa. Introduction: Media as Historical Subjects. In *Always Already New: Media, History, and the Data of Culture*, 1–22. Cambridge, MA: MIT Press, 2006.

GLAAD. "Where We Are on TV, 2015–2016." Los Angeles: GLAAD, 2015. http://www.glaad.org/files/GLAAD-2015-WWAT.pdf.

Glaser, Barney G., and Anselm L. Strauss. *The Discovery of Grounded Theory: Strategies for Qualitative Research*. Chicago: Aldine Publishing Company, 1967.

Glennen, Sharon, and Denise C. DeCoste, eds. *The Handbook of Augmentative and Alternative Communication*. San Diego: Singular Publishing Group, 1997.

Goffman, Erving. *The Presentation of Self in Everyday Life*. New York: Anchor Books, 1959.

Goggin, Gerard. Making Voice Portable: The Early History of the Cell Phone. In *Cell Phone Culture: Mobile Technology in Everyday Life*, 19–40. New York: Routledge, 2006.

Goggin, Gerard. "Disability and the Ethics of Listening." *Continuum: Journal of Media and Cultural Studies* 23, no. 4 (2009): 489–502.

Goggin, Gerard. "Youth Culture and Mobiles." *Mobile Media and Communication* 1, no. 1 (2013): 83–88.

Goggin, Gerard, and Christopher Newell. *Digital Disability: The Social Construction of Disability in New Media*. Lanham, MD: Rowman and Littlefield, 2003.

Goggin, Gerard, and Christopher Newell. Disabling Cell Phones. In *The Cell Phone Reader*, ed. Anandam Kavoori and Noah Arceneaux, 155–172. New York: Peter Lang, 2006.

Gold, Ben, Nelson Morgan, and Dan Ellis. *Speech and Audio Signal Processing: Processing and Perception of Speech and Music*. 2nd ed. Hoboken, NJ: Wiley, 2011.

Golson, Jordan. "iOS 6: Guided Access, Also for Kids and Kiosks." *MacRumors*. September 19, 2012. http://www.macrumors.com/2012/09/19/ios-6-feature-guided-access-also-for-kids-and-kiosks/.

Gosnell, Jessica, John Costello, and Howard Shane. "Using a Clinical Approach to Answer 'What Communication Apps Should We Use?'" *Perspectives on Augmentative and Alternative Communication* 20 (2011): 87–96.

Gray, Mary L. *Out in the Country: Youth, Media, and Queer Visibility in Rural America*. New York: NYU Press, 2009.

Green, Sara E. "'We're Tired, Not Sad': Benefits and Burdens of Mothering a Child with a Disability." *Social Science and Medicine* 64, no. 1 (2007): 150–163.

Gross, Liza. "A Broken Trust: Lessons from the Vaccine-Autism Wars." *PLoS Biology* 7, no. 5 (2009). http://journals.plos.org/plosbiology/article?id=10.1371/journal.pbio.1000114.

Haddon, Leslie. "The Contribution of Domestication Research to In-Home Computing and Media Consumption." *Information Society* 22, no. 4 (2007): 195–203.

Hager, Emily B. "iPad Opens World to a Disabled Boy." *New York Times*, October 31, 2010, B7.

Hager, Emily B. "For Children Who Cannot Speak, a True Voice via Technology." *New York Times*, July 26, 2012, B3.

Halford, Susan, and Mike Savage. "Reconceptualizing Digital Social Inequality." *Information, Communication, and Society* 13, no. 7 (2010): 937–955.

Haller, Beth. *Representing Disability in an Ableist World: Essays on Mass Media.* Louisville, KY: Advocado Press, 2010.

Haller, Beth, Chelsea Temple Jones, Vishaya Naidoo, Art Blaser, and Lindzey Galliford. "iTechnology as Cure or iTechnology as Empowerment: What Do North American News Media Report?" *Disability Studies Quarterly* 36, no. 1 (2016). http://dsq-sds.org/article/view/3857/4208.

Hancock, Ange-Marie. "When Multiplication Doesn't Equal Quick Addition: Examining Intersectionality as a Research Paradigm." *Perspectives on Politics* 5, no. 1 (2007): 63–79.

Hankins, Thomas, and Robert Silverman. *Vox Mechanica*: The History of Speaking Machines. In *Instruments and the Imagination*, 178–220. Princeton, NJ: Princeton University Press, 1995.

Haraway, Donna J. *Simians, Cyborgs, and Women: The Re-Invention of Nature.* New York: Routledge, 1991.

Haraway, Donna J. Able Bodies and Companion Species. In *When Species Meet*, 161–179. Minneapolis: University of Minnesota Press, 2007.

Hayles, N. Katherine. *How We Became Posthuman: Virtual Bodies in Cybernetics, Literature, and Informatics*. Chicago: University of Chicago Press, 1999.

Hays, Sharon. "Structure and Agency and the Sticky Problem of Culture." *Sociological Theory* 12, no. 1 (1994): 57–72.

Hays, Sharon. *The Cultural Contradictions of Motherhood.* New Haven, CT: Yale University Press, 1996.

Hemsley, Bronwyn, Stephen Dann, Stuart Palmer, Meredith Allan, and Susan Balandin. "'We Definitely Need an Audience': Experiences of Twitter, Twitter Networks, and Tweet Content in Adults with Severe Communication Disabilities Who Use Augmentative and Alternative Communication (AAC)." *Disability and Rehabilitation* 37, no. 17 (2015): 1531–1542.

Hendren, Sara. "All Technology Is Assistive Technology." *Abler*, September 17, 2013. http://ablersite.org/2013/09/17/all-technology-is-assistive-technology/

Herr-Stephenson, Becky, and Meryl Alper (with Erin Reilly and Henry Jenkins). *T Is for Transmedia: Learning through Transmedia Play*. Los Angeles: USC Annenberg Innovation Lab; New York: The Joan Ganz Cooney Center at Sesame Workshop, 2013.

Hershberger, Dave. "Mobile Technology and AAC Apps from an AAC Developer's Perspective." *Perspectives on Augmentative and Alternative Communication* 20, no. 1 (2011): 28–33.

Higashida, Naoki. *The Reason I Jump: The Inner Voice of a Thirteen-Year-Old Boy with Autism*. New York: Random House, 2013.

Higginbotham, Jeff. "Problems and Solutions in Data Logging." *Alternatively Speaking* 5, no. 2 (2000): 4–5.

Higginbotham, Jeff, and Steve Jacobs. "The Future of the Android Operating System for Augmentative and Alternative Communication." *Perspectives on Augmentative and Alternative Communication* 20, no. 2 (2011): 52–56.

Hill, Katya, Barry Romich, and Snoopi J. Botten. Rights and Privacy in AAC Evidence-Based Clinical Practice. In *Proceedings of the Technology and Persons with Disabilities Conference*. Northridge: California State University, 2002. http://www.csun.edu/~hfdss006/conf/2002/proceedings/247.htm.

HillaryClinton.com. "Factsheets: Hillary Clinton's Plan to Support Children, Youth, and Adults Living with Autism and their Families." Posted January 5, 2016. https://www.hillaryclinton.com/briefing/factsheets/2016/01/05/hillary-clintons-plan-to-support-children-youth-and-adults-living-with-autism-and-their-families/.

Hjorth, Larissa. "Odours of Mobility: Mobile Phones and Japanese Cute Culture in the Asia-Pacific." *Journal of Intercultural Studies* 26, no. 4 (2005): 39–55.

Hjorth, Larissa, and Sun Sun Lim. "Mobile Intimacy in an Age of Affective Mobile Media." *Feminist Media Studies* 12, no. 4 (2012): 477–484.

Hochschild, Arlie R. *The Managed Heart: Commercialization of Human Feeling*. Berkeley: University of California Press, 1983.

Hochschild, Arlie R. *The Second Shift: Working Parents and the Revolution at Home*. Berkeley: University of California Press, 1989.

Holmes, Seth M. *Fresh Fruit, Broken Bodies: Migrant Farmworkers in the United States*. Berkeley: University of California Press, 2013.

Hoover, Stewart M., Lynn Schofield Clark, and Diane Alters. *Media, Home, and Family*. New York: Routledge, 2004.

Horst, Heather A., Becky Herr-Stephenson, and Laura Robinson. Media Ecologies. In *Hanging Out, Messing Around, and Geeking Out: Kids Living and Learning with New Media*, ed. Mizuko Ito et al., 29–78. Cambridge, MA: MIT Press, 2010.

Horst, Heather A., and Daniel Miller. *The Cell Phone: An Anthropology of Communication*. New York: Berg, 2006.

Houtrow, Amy J., Kandyce Larson, Lynn M. Olson, Paul W. Newacheck, and Neal Halfon. "Changing Trends of Childhood Disability, 2001–2011." *Pediatrics* 134 (2014): 530–538.

Howery, Kathy L. Look. Speech-Generating Devices in the Lives of Young People with Severe Speech Impairment: What Does the Non-Speaking Child Say? In *Efficacy of Assistive Technology Interventions*, ed. Dave L. Edyburn, 79–109. Bingley, UK: Emerald, 2015.

Hulko, Wendy. "The Time- and Context-Contingent Nature of Intersectionality and Interlocking Oppressions." *Affilia: Journal of Women and Social Work* 24, no. 1 (2009): 44–55.

Hutchby, Ian. *Conversation and Technology: From the Telephone to the Internet*. Cambridge, UK: Polity, 2001.

Hynan, Amanda, Juliet Goldbart, and Janice Murray. "A Grounded Theory of Internet and Social Media Use by Young People Who Use Augmentative and Alternative Communication (AAC)." *Disability and Rehabilitation* 37, no. 17 (2015): 1559–1575.

Hynan, Amanda, Janice Murray, and Juliet Goldbart. "'Happy and Excited': Perceptions of Using Digital Technology and Social Media by Young People Who Use Augmentative and Alternative Communication." *Child Language Teaching and Therapy* 30, no. 2 (2014): 175–186.

Ihde, Don. Embodying Hearing Devices: Digitalization. In *Listening and Voice: Phenomenologies of Sound*, 243–250. 2nd ed. Albany: State University of New York Press, 2007.

Ito, Mizuko. Mobilizing the Imagination in Everyday Play: The Case of Japanese Media Mixes. In *The International Handbook of Children, Media, and Culture*, ed. Kirsten Drotner and Sonia Livingstone, 397–412. London: Sage, 2008.

Ito, Mizuko. *Engineering Play: A Cultural History of Children's Software*. Cambridge, MA: MIT Press, 2009.

Ito, Mizuko, Sonja Baumer, Matteo Bittanti, danah boyd, Rachel Cody, Becky Herr Stephenson, Heather A. Horst, Patricia G. Lange, Dilan Mahendran, Katynka Z. Martínez, C. J. Pascoe, Dan Perkel, Laura Robinson, Christo Sims and Lisa Tripp. *Hanging Out, Messing Around, and Geeking Out: Kids Living and Learning with New Media*. Cambridge, MA: MIT Press, 2009.

Mizuko, Ito, Kris Gutiérrez, Sonia Livingstone, Bill Penuel, Jean Rhodes, Katie Salen, Juliet Schor, Julian Sefton-Green, and S. Craig Watkins. *Connected Learning: An Agenda for Research and Design*. Irvine, CA: Digital Media and Learning Research Hub, 2013.

Ito, Mizuko, Daisuke Okabe, and Misa Matsuda, eds. *Personal, Portable, Pedestrian: Mobile Phones in Japanese Life*. Cambridge, MA: MIT Press, 2005.

Jack, Jordynn. *Autism and Gender: From Refrigerator Mothers to Computer Geeks*. Champaign: University of Illinois Press, 2014.

James, Allison, Chris Jenks, and Alan Prout. *Theorizing Childhood*. Oxford: Polity Press, 1998.

Jenkins, Henry. Introduction: "Worship at the Altar of Convergence": A New Paradigm for Understanding Media Change. In *Convergence Culture: Where Old and New Media Collide*, 1–24. New York: NYU Press, 2006.

Jenkins, Henry. The Guiding Spirit and the Powers That Be: A Response to Suzanne Scott. In *The Participatory Cultures Handbook*, ed. Aaron Delwiche and Jennifer Jacobs Henderson, 53–58. New York: Routledge, 2013.

Jenkins, Henry, with Ravi Purushotma, Margaret Weigel, Katie Clinton, and Alice J. Robison. *Confronting the Challenges of Participatory Culture: Media Education for the 21st Century*. Chicago: John D. and Catherine T. MacArthur Foundation, 2006.

Jonze, Spike. *Her*. Burbank, CA: Warner Bros., 2013.

Jordan, Amy B. "Social Class, Temporal Orientation, and Mass Media Use within the Family System." *Critical Studies in Media Communication* 9 (1992): 374–386.

Jordan, Chloe J. "Evolution of Autism Support and Understanding via the World Wide Web." *Intellectual and Developmental Disabilities* 48, no. 3 (2009): 220–227.

Kafai, Yasmin B., Melissa S. Cook, and Deborah A. Fields. "'Blacks Deserve Bodies Too!' Design and Discussion about Diversity and Race in a Tween Virtual World." *Games and Culture* 5, no. 1 (2010): 43–63.

Kafer, Alison. *Feminist, Queer, Crip*. Bloomington: Indiana University Press, 2013.

Kang, Cecilia, and Eric Lichtblau. "F.B.I. Error Locked San Bernardino Attacker's iPhone." *New York Times*, March 1, 2016. http://www.nytimes.com/2016/03/02/technology/apple-and-fbi-face-off-before-house-judiciary-committee.html.

Kanner, Leo. "Problems of Nosology and Psychodynamics of Early Infantile Autism." *American Journal of Orthopsychiatry* 19, no. 3 (1949): 416–426.

Kapp, Steven K., Kristen Gillespie-Lynch, Lauren E. Sherman, and Ted Hutman. "Deficit, Difference, or Both? Autism and Neurodiversity." *Developmental Psychology* 49 (1) (2013): 59–71.

Kasari, Connie, Ann Kaiser, Kelly Goods, Jennifer Nietfeld, Pamela Mathy, Rebecca Landa, Susan Murphy, and Daniel Almirall. "Communication Interventions for Minimally Verbal Children with Autism: A Sequential Multiple Assignment Randomized Trial." *Journal of the American Academy of Child and Adolescent Psychiatry* 53, no. 6 (2014): 635–646.

Katz, James E., and Mark Aakhus, eds. *Perpetual Contact: Mobile Communication, Private Talk, Public Performance*. Cambridge, UK: Cambridge University Press, 2002.

Katz, James E., and Satomi Sugiyama. "Mobile Phones as Fashion Statements: Evidence from Student Surveys in the US and Japan." *New Media and Society* 8, no. 2 (2006): 321–337.

Katz, Vikki S. "How Children of Immigrants Use Media to Connect Their Families to the Community." *Journal of Children and Media* 4, no. 3 (2010): 298–315.

Katz, Vikki S. *Kids in the Middle: How Children of Immigrants Negotiate Community Interactions for Their Families*. New Brunswick, NJ: Rutgers University Press, 2014.

Kedar, Ido. *Ido in Autismland: Climbing out of Autism's Silent Prison*. Self-published, 2012.

Keilty, Bonnie, and Kristin M. Galvin. "Physical and Social Adaptations of Families to Promote Learning in Everyday Experiences." *Topics in Early Childhood Special Education* 26, no. 4 (2006): 219–233.

Kendon, Adam. *Gesture: Visible Action as Utterance*. Cambridge, UK: Cambridge University Press, 2004.

Kennedy, Kathleen, Jon Kilik, and Julian Schnabel. *The Diving Bell and the Butterfly*. New York: Miramax Films, 2007.

Kevorkian, Martin. Integrated Circuits. In *Color Monitors: The Black Face of Technology in America*, 74–114. Ithaca, NY: Cornell University Press, 2006.

Kirkpatrick, Bill. "'A Blessed Boon': Radio, Disability, Governmentality, and the Discourse of the 'Shut-In,' 1920–1930." *Critical Studies in Media Communication* 29, no. 3 (2012): 165–184.

Klatt, Dennis H. "Review of Text-to-Speech Conversion for English." *Journal of the Acoustical Society of America* 82, no. 3 (1987): 737–793.

Komulainen, Sirkka. "The Ambiguity of the Child's 'Voice' in Social Research." *Childhood* 14, no. 1 (2007): 11–28.

Kress, Gunther. *Before Writing: Rethinking the Paths to Literacy*. New York: Routledge, 1997.

Kumparak, Greg. "iTunes Rewind Highlights the Best Apps of 2009." *TechCrunch*, December 8, 2009. http://techcrunch.com/2009/12/08/itunes-rewind-highlights-the-best-apps-of-2009/.

Kvasny, Lynette. "Cultural (Re)production of Digital Inequality in a U.S. Community Technology Initiative." *Information, Communication, and Society* 9, no. 2 (2006): 160–181.

Lacey, Kate. *Listening Publics: The Politics and Experience of Listening in the Media Age*. Cambridge, UK: Polity Press, 2013.

Lally, Elaine. *At Home with Computers*. New York: Berg, 2002.

Lamont, Michèle, and Annette Lareau. "Cultural Capital: Allusions, Gaps, and Glissandos in Recent Theoretical Developments." *Sociological Theory* 6, no. 2 (1988): 153–168.

Landsman, Gail. "Real Motherhood," Class, and Children with Disabilities. In *Ideologies and Technologies of Motherhood: Race, Class, Sexuality, Nationalism*, ed. Heléna Ragoné and France W. Twine, 169–187. New York: Routledge, 2000.

Landsman, Gail. *Reconstructing Motherhood and Disability in the Age of "Perfect" Babies*. New York: Routledge, 2009.

Langan, Mary. "Parental Voices and Controversies in Autism." *Disability and Society* 26, no. 2 (2011): 193–205.

Lankshear, Colin, and Michele Knobel. *New Literacies: Everyday Practices and Classroom Learning*. 3rd ed. New York: Open University Press, 2011.

Lareau, Annette. *Home Advantage: Social Class and Parental Intervention in Elementary Education*. Oxford: Rowman and Littlefield, 2000.

Lareau, Annette. *Unequal Childhoods: Class, Race, and Family Life*. Berkeley: University of California Press, 2003.

Lasker, Joanne, and Jan Bedrosian. "Promoting Acceptance of Augmentative and Alternative Communication by Adults with Acquired Communication Disorders." *Augmentative and Alternative Communication* 17, no. 3 (2011): 141–153.

Latour, Bruno. Technology Is Society Made Durable. In *A Sociology of Monsters: Essays on Power, Technology, and Domination*, ed. John Law, 103–131. London: Routledge, 1991.

Lee, Jung-Sook, and Natasha K. Bowen. "Parent Involvement, Cultural Capital, and the Achievement Gap among Elementary School Children." *American Educational Research Journal* 43, no. 2 (2006): 193–218.

Lievrouw, Leah A. Materiality and Media in Communication and Technology Studies: An Unfinished Project. In *Media Technologies: Essays on Communication, Materiality, and Society*, ed. Tarleton Gillespie, Pablo J. Boczkowski, and Kirsten A. Foot, 21–52. Cambridge, MA: MIT Press, 2014.

Lievrouw, Leah A., and Sonia Livingstone. Introduction to *Handbook of New Media: Social Shaping and Social Consequences*, ed. Leah A. Lievrouw and Sonia Livingstone, 1–14. London: Sage, 2006.

Light, J., K. D. Drager, and J. G. Nemser. "Enhancing the Appeal of AAC Technologies for Young Children: Lessons from the Toy Manufacturers." *Augmentative and Alternative Communication* 20 (2004): 137–149.

Light, Janice, and David McNaughton. "Supporting the Communication, Language, and Literacy Development of Children with Complex Communication Needs: State of the Science and Future Research Priorities." *Assistive Technology: The Official Journal of RESNA* 24, no. 1 (2012): 34–44.

Light, Janice, and David McNaughton. "Communicative Competence for Individuals Who Require Augmentative and Alternative Communication: A New Definition for a New Era of Communication?" *Augmentative and Alternative Communication* 30, no. 1 (2014): 1–18.

Lin, Sue C., Stella M. Yu, and Robin L. Harwood. "Autism Spectrum Disorders and Developmental Disabilities in Children from Immigrant Families in the United States." *Pediatrics* 13 supplement 2 (2012): S191–S197.

Lindlof, Thomas R., and Bryan C. Taylor. *Qualitative Communication Research Methods*. 2nd ed. Thousand Oaks, CA: Sage, 2002.

Lindsay, David. "Talking Head." *American Heritage of Invention and Technology* 13, no. 1 (1997): 57–63.

Linton, Simi. *Claiming Disability: Knowledge and Identity*. New York: NYU Press, 1998.

Liss, Miriam, Celine Saulnier, Deborah Fein, and Marcel Kinsbourne. "Sensory and Attention Abnormalities in Autistic Spectrum Disorders." *Autism* 10, no. 2 (2006): 155–172.

Livingstone, Sonia. *Young People and New Media: Childhood and the Changing Media Environment*. London: Sage, 2002.

Livingstone, Sonia, and Moira Bovill. *Families and the Internet: An Observational Study of Children and Young People's Internet Use* . London: Media@LSE, 2001.

Livingstone, Sonia, and Ellen Helsper. "Gradations in Digital Inclusion: Children, Young People, and the Digital Divide." *New Media and Society* 9, no. 4 (2007): 671–696.

Lorah, Elizabeth R., Ashley Parnell, Peggy Schaefer Whitby, and Donald Hantula. "A Systematic Review of Tablet Computers and Portable Media Players as Speech Generating Devices for Individuals with Autism Spectrum Disorder." *Journal of Autism and Developmental Disorders* 45 (2015): 3792–3804.

Lovato, Silvia, and Anne Marie Piper. "'Siri, Is This You?': Understanding Young Children's Interactions with Voice Input Systems." In *Proceedings of the Annual Conference on Interaction Design and Children (IDC)*, 335–338. New York: ACM Press, 2015.

Lull, James. *Inside Family Viewing: Ethnographic Research on Television's Audiences.* New York: Routledge, 1990.

Mackintosh, Virginia H., Barbara J. Myers, and Robin P. Goin-Kochel. "Sources of Information and Support Used by Parents of Children with Autism Spectrum Disorders." *Journal on Developmental Disabilities* 12, no. 4 (2005): 1–51.

MacMullin, Jennifer A., Yona Lunsky, and Jonathan A. Weiss. "Plugged In: Electronics Use in Youth and Young Adults with Autism Spectrum Disorder." *Autism* 20, no. 1 (2016): 45–54.

Macnamara, Jim. "Beyond Voice: Audience-Making and the Work and Architecture of Listening as New Media Literacies." *Continuum: Journal of Media and Cultural Studies* 27, no. 1 (2013): 160–175.

Mankoff, Jennifer, Gillian R. Hayes, and Devva Kasnitz. "Disability Studies as a Source of Critical Inquiry for the Field of Assistive Technology." In *Proceedings of the SIGACCESS Conference on Computers and Accessibility*, 3–10. New York: ACM Press, 2010.

Marsh, Jackie, ed. *Popular Culture, New Media, and Digital Literacy in Early Childhood.* New York: Routledge Falmer, 2005.

Marshall, Julie, and Juliet Goldbart. "'Communication Is Everything I Think': Parenting a Child Who Needs Augmentative and Alternative Communication (AAC)." *International Journal of Language and Communication Disorders* 43, no. 1 (2008): 77–98.

Mattingly, Cheryl. "Becoming Buzz Lightyear and Other Clinical Tales: Indigenizing Disney in a World of Disability." *Folk* 45 (2003): 9–32.

Mattingly, Cheryl. "Pocahontas Goes to the Clinic: Popular Culture as Lingua Franca in a Cultural Borderland." *American Anthropologist* 108, no. 3 (2006): 494–501.

Maul, Christine A., and George H. S. Singer. "'Just Good Different Things': Specific Accommodations Families Make to Positively Adapt to Their Children with Developmental Disabilities." *Topics in Early Childhood Special Education* 29, no. 3 (2009): 155–170.

Mazurek, Micah O., and Christopher R. Engelhardt. "Video Game Use and Problem Behaviors in Boys with Autism Spectrum Disorders." *Research in Autism Spectrum Disorders* 7, no. 2 (2013): 316–324.

Mazurek, Micah O., and Christopher R. Engelhardt. "Video Game Use in Boys with Autism Spectrum Disorder, ADHD, or Typical Development." *Pediatrics* 132, no. 2 (2013): 260–266.

McCall, Leslie. "The Complexity of Intersectionality." *Signs* 30, no. 3 (2005): 1771–1800.

McCord, Shannon M., and Gloria Soto. "Perceptions of AAC: An Ethnographic Investigation of Mexican-American Families." *Augmentative and Alternative Communication* 20, no. 4 (2004): 209–227.

McIntosh, Peggy. "White Privilege and Male Privilege: A Personal Account of Coming to See Correspondences through Work in Women's Studies." Working paper, Wellesley College Center for Research on Women, Wellesley, MA, 1988.

McLellan, Dennis. "Electronic Help for the Handicapped: The Voiceless Break Their Silence." *Los Angeles Times*, December 29, 1980, B1.

McLuhan, Marshall. *Understanding Media: The Extensions of Man*. Reprint ed. Cambridge, MA: MIT Press, 1994.

McNaughton, David, and Janice Light. "The iPad and Mobile Technology Revolution: Benefits and Challenges for Individuals Who Require Augmentative and Alternative Communication." *Augmentative and Alternative Communication* 29 (2013): 107–116.

McNaughton, David, Tracy Rackensperger, Elizabeth Benedek-Wood, Carole Krezman, Michael B. Williams, and Janice Light. "'A Child Needs to Be Given a Chance to Succeed': Parents of Individuals Who Use AAC Describe the Benefits and Challenges of Learning AAC Technologies." *Augmentative and Alternative Communication* 24, no. 1 (2008): 43–55.

McRuer, Robert. *Crip Theory: Cultural Signs of Queerness and Disability*. New York: NYU Press, 2006.

Meder, Allison M., and Jane R. Wegner. "iPads, Mobile Technologies, and Communication Applications: A Survey of Family Wants, Needs, and Preferences." *Augmentative and Alternative Communication* 31, no. 1 (2015): 27–36.

Mialet, Hélène. *Hawking Incorporated: Stephen Hawking and the Anthropology of the Knowing Subject.* Chicago: University of Chicago Press, 2012.

Microsoft. "Microsoft 2014 Super Bowl Commercial: Empowering." YouTube video. Posted February 2, 2014. https://www.youtube.com/watch?v=qaOvHKG0Tio.

Mills, C. Wright. *The Sociological Imagination.* New York: Oxford University Press, 1959.

Mills, Mara. "When Mobile Communication Technologies Were New." *Endeavor* 33, no. 4 (2009): 140–146.

Mills, Mara. "Media and Prosthesis: The Vocoder, the Artificial Larynx, and the History of Signal Processing." *Qui Parle: Critical Humanities and Social Sciences* 21, no. 1 (2012): 107–149.

Mitra, Ananda, and Eric Watts. "Theorizing Cyberspace: The Idea of Voice Applied to the Internet Discourse." *New Media and Society* 4, no. 4 (2002): 479–498.

Mittell, Jason. "The Cultural Power of an Anti-Television Metaphor: Questioning the 'Plug-in Drug' and a TV-Free America." *Television and New Media* 1, no. 2 (2000): 215–238.

Morley, David. *Family Television: Cultural Power and Domestic Leisure.* New York: Routledge, 1988.

Morley, David. *Home Territories: Media, Mobility, and Identity.* London: Routledge, 2000.

Morris, Merrie R. "Social Networking Site Use by Mothers of Young Children." In *Proceedings of the Annual Conference on Computer Supported Cooperative Work (CSCW),* 1272–1282. New York: ACM Press, 2014.

Moser, Ingunn. "Disability and the Promises of Technology: Technology, Subjectivity, and Embodiment within an Order of the Normal." *Information, Communication, and Society* 9, no. 3 (2006): 373–395.

Moser, Ingunn. "Sociotechnical Practices and Difference: On the Interferences between Disability, Gender, and Class." *Science, Technology, and Human Values* 31, no. 5 (2006): 537–564.

Moser, Ingunn, and John Law. "Making Voices": New Media Technologies, Disabilities, and Articulation. In *Digital Media Revisited: Theoretical and Conceptual Innovation in Digital Domains,* ed. Gunnar Liestøl, Terje Rasmussen, and Andrew Morrison, 491–520. Cambridge, MA: MIT Press, 2003.

Nagy, Peter, and Gina Neff. "Imagined Affordance: Reconstructing a Keyword for Communication Theory." *Social Media + Society* 1, no. 2 (2015): 1–9.

Nakamura, Lisa. *Cybertypes: Race, Ethnicity, and Identity on the Internet.* New York: Routledge, 2002.

Nally, Brenda, Bob Houlton, and Sue Ralph. "Researches in Brief: The Management of Television and Video by Parents of Children with Autism." *Autism* 4, no. 3 (2000): 331–338.

Naraian, Srikala. "Disentangling the Social Threads within a Communicative Environment: A Cacophonous Tale of Alternative and Augmentative Communication (AAC)." *European Journal of Special Needs Education* 25, no. 3 (2010): 253–267.

Nash, Jennifer C. "Re-Thinking Intersectionality." *Feminist Review* 89 (2008): 1–15.

Nass, Clifford, and Scott Brave. *Wired for Speech: How Voice Activates and Advances the Human–Computer Relationship.* Cambridge, MA: MIT Press, 2005.

National Academies of Sciences, Engineering, and Medicine. *Speech and Language Disorders in Children: Implications for the Social Security Administration's Supplemental Security Income Program.* Washington, DC: National Academies of Sciences, Engineering, and Medicine, 2016.

National Joint Committee for the Communication Needs of Persons with Severe Disabilities. *Guidelines for Meeting the Communication Needs of Persons with Severe Disabilities*, 1992. http://www.asha.org/policy/GL1992-00201/.

Newman, Judith. "To Siri with Love: How One Boy with Autism Became BFF with Apple's Siri." *New York Times*, October 17, 2014. http://www.nytimes.com/2014/10/19/fashion/how-apples-siri-became-one-autistic-boys-bff.html.

Newman, Michael Z. "New Media, Young Audiences, and Discourses of Attention: From Sesame Street to 'Snack Culture.'" *Media, Culture, and Society* 32, no. 4 (2010): 581–596.

Niemeijer, David. "The State of AAC in English-Speaking Countries: First Results from the Survey." *AssistiveWare* (blog), n.d. http://www.assistiveware.com/state-aac-english-speaking-countries-first-results-survey.

Niemeijer, David, Anne M. Donnellan, and Jodi A. Robledo. *Taking the Pulse of Augmentative and Alternative Communication on iOS.* Amsterdam: AssistiveWare, 2012. http://www.assistiveware.com/taking-pulse-augmentative-and-alternative-communication-ios.

Norman, Don. *The Design of Everyday Things.* New York: Doubleday, 1988.

Norrelgen, Fritjof, Elisabeth Fernell, Mats Eriksson, Åsa Hedvall, Clara Persson, Maria Sjölin, Christopher Gillberg, and Liselotte Kjellmer. "Children with Autism Spectrum Disorders Who Do Not Develop Phrase Speech in the Preschool Years." *Autism* 19, no. 8 (2015): 934–943.

Norris, Pippa, and Dieter Zinnbauer. "Giving Voice to the Voiceless: Good Governance, Human Development, and Mass Communications." Human Development Report Office Occasional Paper 2002/2011. New York: United Nations Development Programme, 2002.

Olesen, Virginia. Feminist Qualitative Research in the Millennium's First Decade. In *The Sage Handbook of Qualitative Research*, ed. Norman K. Denzin and Yvonna S. Lincoln, 129–146. Thousand Oaks, CA: Sage, 2011.

Olive, Joseph P. "The Talking Computer": Text to Speech Synthesis. In *HAL's Legacy: 2001's Computer as Dream and Reality*, ed. David Stork, 101–129. Cambridge, MA: MIT Press, 1997.

Oliver, Michael. *The Politics of Disablement.* Basingstoke, UK: Macmillan, 1990.

Ong, Walter J. *Interfaces of the Word.* Ithaca, NY: Cornell University Press, 1977.

Ong, Walter J. *Orality and Literacy: The Technologizing of the Word.* New York: Methuen, 1982.

Ong-Dean, Colin. "Reconsidering the Social Location of the Medical Model: An Examination of Disability in Parenting Literature." *Journal of Medical Humanities* 26, no. 2–3 (2005): 141–158.

Ong-Dean, Colin. *Distinguishing Disability: Parents, Privilege, and Special Education.* Chicago: University of Chicago Press, 2009.

Ott, Katherine. The Sum of Its Parts: An Introduction to Modern Histories of Prosthetics. In *Artificial Parts, Practical Lives: Modern Histories of Prosthetics*, ed. Katherine Ott, David Serlin, and Stephen Mihm, 1–42. New York: NYU Press, 2002.

Parish, Susan L., Roderick A. Rose, Sarah Dababnah, Joan Yoo, and Shawn A. Cassiman. "State-Level Income Inequality and Family Burden of US Families Raising Children with Special Health Care Needs." *Social Science and Medicine* 74, no. 3 (2012): 399–407.

Park, Haeyoun. "Paris Attacks Intensify Debate over How Many Syrian Refugees to Allow into the U.S." *New York Times*, November 19, 2015. http://www.nytimes.com/interactive/2015/10/21/us/where-syrian-refugees-are-in-the-united-states.html.

Patterson, Joan M. "Integrating Family Resilience and Family Stress Theory." *Journal of Marriage and the Family* 64, no. 2 (2002): 349–360.

Peppler, Kylie A., and Mark Warschauer. "Uncovering Literacies, Disrupting Stereotypes: Examining the (Dis)abilities of a Child Learning to Computer Program and Read." *International Journal of Learning and Media* 3, no. 3 (2012): 15–41.

Peters, John Durham. *Speaking into the Air: A History of the Idea of Communication.* Chicago: University of Chicago Press, 1999.

Peters, John Durham. "The Voice between Phenomenology, Media, and Religion." *Glimpse: The Journal of the Society for Phenomenology and Media* 6 (2005): 1–10.

Petrick, Elizabeth. *Making Computers Accessible: Disability Rights and Digital Technology*. Baltimore: Johns Hopkins University Press, 2015.

Pistorius, Martin. *Ghost Boy: The Miraculous Escape of a Misdiagnosed Boy Trapped Inside His Own Body*. Nashville, TN: Nelson Books, 2013.

Pitaru, Amit. E Is for Everyone: The Case for Inclusive Game Design. In *The Ecology of Games: Connecting Youth, Games, and Learning*, ed. Katie Salen, 67–88. Cambridge, MA: MIT Press, 2008.

Plantin, Lars, and Kristian Daneback. "Parenthood, Information, and Support on the Internet: A Literature Review of Research on Parents and Professionals Online." *BMC Family Practice* 10, no. 34 (2009). http://www.biomedcentral.com/1471-2296/10/34.

Pugh, Allison J. *Longing and Belonging: Parents, Children, and Consumer Culture*. Berkeley: University of California Press, 2009.

Pullin, Graham. *Design Meets Disability*. Cambridge, MA: MIT Press, 2011.

Pullin, Graham. Expression Meets Information. In *Design Meets Disability*, 155–179. Cambridge, MA: MIT Press, 2011.

Pullin, Graham, and Shannon Hennig. "17 Ways to Say Yes: Toward Nuanced Tone of Voice in AAC and Speech Technology." *Augmentative and Alternative Communication* 31, no. 2 (2015): 170–180.

Radway, Janice. *Reading the Romance: Women, Patriarchy, and Popular Literature*. Chapel Hill: University of North Carolina Press, 1984.

Ragin, Charles C., and Howard S. Becker, eds. *What Is a Case? Exploring the Foundations of Social Inquiry*. Cambridge, UK: Cambridge University Press, 1992.

Rakow, Lana F., and Vija Navarro. "Remote Mothering and the Parallel Shift: Women Meet the Cellular Phone." *Critical Studies in Mass Communication* 10, no. 2 (1993): 144–157.

Rapp, Rayna, and Faye Ginsburg. "Enabling Disability: Rewriting Kinship, Reimagining Citizenship." *Public Culture* 13, no. 3 (2001): 533–556.

Rapp, Rayna, and Faye Ginsburg. "Reverberations: Disability and the New Kinship Imaginary." *Anthropological Quarterly* 84, no. 2 (2011): 379–410.

Ratto, Matt, and Megan Boler, eds. *DIY Citizenship: Critical Making and Social Media*. Cambridge, MA: MIT Press, 2014.

Reed, Lori. "Domesticating the Personal Computer: The Mainstreaming of a New Technology and the Cultural Management of a Widespread Technophobia, 1964–." *Critical Studies in Media Communication* 17, no. 2 (2000): 159–185.

Reeves, Joshua. "Automatic for the People: The Automation of Communicative Labor." *Communication and Critical/Cultural Studies*. Forthcoming.

Reinke, Jennifer S., and Catherine A. Solheim. "Online Social Support Experiences of Mothers of Children with Autism Spectrum Disorder." *Journal of Child and Family Studies* 24, no. 8 (2015): 2364–2373.

Reno, Joshua. "Technically Speaking: On Equipping and Evaluating 'Unnatural' Language Learners." *American Anthropologist* 114, no. 3 (2012): 406–419.

Resnick, Mitchel, Amy Bruckman, and Fred Martin. "Pianos Not Stereos: Creating Computational Construction Kits." *Interaction* 3, no. 6 (1996): 40–50.

Ribak, Rivka. "Children and New Media: Reflections on the Ampersand." *Journal of Children and Media* 1, no. 1 (2007): 68–76.

Rideout, Victoria. *Zero to Eight: Children's Media Use in America 2013*. San Francisco: Common Sense Media, 2013.

Rideout, Victoria. *Learning at Home: Families' Educational Media Use in America*. New York: Joan Ganz Cooney Center at Sesame Workshop, 2014.

Rideout, Victoria, and Vikki S. Katz. *Opportunity for All? Technology and Learning in Lower-Income Families*. New York: Joan Ganz Cooney Center at Sesame Workshop, 2016.

Ripat, Jacquie, and Roberta Woodgate. "The Intersection of Culture, Disability, and Assistive Technology." *Disability and Rehabilitation: Assistive Technology* 6, no. 2 (2011): 87–96.

Rispoli, Mandy J., Jessica H. Franco, Larah van der Meer, Russell Lang, and Síglia P. H. Camargo. "The Use of Speech Generating Devices in Communication Interventions for Individuals with Developmental Disabilities: A Review of the Literature." *Developmental Neurorehabilitation* 13, no. 4 (2010): 276–293.

Robertson, Scott M. "Neurodiversity, Quality of Life, and Autistic Adults: Shifting Research and Professional Focuses onto Real-Life Challenges." *Disability Studies Quarterly* 30, no. 1 (2010). http://dsq-sds.org/article/view/1069/1234.

Robinson, Muriel, and Margaret Mackey. Film and Television. In *Handbook of Early Childhood Literacy*, ed. Joanne Larson and Jackie Marsh, 223–250. 2nd ed. Thousand Oaks, CA: Sage, 2013.

Robinson, Muriel, and Bernardo Turnbull. Verónica: An Asset Model of Becoming Literate. In *Popular Culture, New Media, and Digital Literacy in Early Childhood*, ed. Jackie Marsh, 51–72. New York: Routledge Falmer, 2005.

Rodino-Colocino, Michelle. "Selling Women on PDAs from 'Simply Palm' to 'Audrey': How Moore's Law Met Parkinson's Law in the Kitchen." *Critical Studies in Media Communication* 23, no. 5 (2006): 375–390.

Rogoff, Barbara. *The Cultural Nature of Human Development*. New York: Oxford University Press, 2003.

Rose, David H., and Anne Meyer. *Teaching Every Student in the Digital Age: Universal Design for Learning*. Alexandria, VA: Association for Supervision and Curriculum Development, 2002.

Rosenberg, Steven A. "Tradition Gets an Update: Technology Helps Autistic 12-Year-Old Find a Voice for His Bar Mitzvah." *Boston Globe*, March 4, 2012, B1.

Roth, Lorna. "Looking at Shirley, the Ultimate Norm: Colour Balance, Image Technologies, and Cognitive Equity." *Canadian Journal of Communication* 34, no. 1 (2009): 111–136.

Rummel-Hudson, Robert. "A Revolution at Their Fingertips." *Perspectives on Augmentative and Alternative Communication* 20, no. 1 (2011): 19–23.

Rutkin, Aviva. "A Voice to Call Your Own." *New Scientist* 221, no. 2961 (2014): 22.

Ryan, Sara, and Katherine Runswick-Cole. "Repositioning Mothers: Mothers, Disabled Children, and Disability Studies." *Disability and Society* 23, no. 3 (2008): 199–210.

Samuels, Ellen. "Critical Divides: Judith Butler's Body Theory and the Question of Disability." *NWSA Journal* 14, no. 3 (2002): 58–76.

Satterfield, Lisa. "The Latest on Medicare Coverage of SGDs." *ASHA Leader* 20 (2015): 28–29.

Sawer, Patrick. "Hawking's Fear Cuts Pose Threat for Disabled Students." *Telegraph*, May 30, 2015. http://www.telegraph.co.uk/news/science/stephen-hawking/11640763/Hawkings-fear-cuts-pose-threat-for-disabled-students.html.

Scherer, Marcia J. *Living in the State of Stuck: How Assistive Technology Impacts the Lives of People with Disabilities*. 4th ed. Cambridge, MA: Brookline Books, 2005.

Schweik, Susan M. *The Ugly Laws: Disability in Public*. New York: NYU Press, 2009.

Seelman, Katherine. "Assistive Technology Policy: A Road to Independence for Individuals with Disabilities." *Journal of Social Issues* 29, no. 2 (1993): 115–136.

Seiter, Ellen. *Sold Separately: Children and Parents in Consumer Culture*. New Brunswick, NJ: Rutgers University Press, 1995.

Seiter, Ellen. Power Rangers at Preschool: Negotiating Media in Child Care Settings. In *Kids' Media Culture*, ed. Marsha Kinder, 239–262. Durham, NC: Duke University Press, 1999.

Seiter, Ellen. *The Internet Playground: Children's Access, Entertainment, and Mis-Education*. New York: Peter Lang, 2005.

Seiter, Ellen. Practicing at Home: Computers, Pianos, and Cultural Capital. In *Digital Youth, Innovation, and the Unexpected*, ed. Tara McPherson, 27–52. Cambridge, MA: MIT Press, 2008.

Sen, Amartya. *Development as Freedom*. New York: Anchor Books, 1999.

Sennott, Samuel. "An Introduction to the Special Issue on New Mobile AAC Technologies." *Perspectives on Augmentative and Alternative Communication* 20 (2011): 3–6.

Shakespeare, Tom. *Disability Rights and Wrongs Revisited*. New York: Routledge, 2013.

Shapiro, Joseph P. *No Pity: People with Disabilities Forging a New Civil Rights Movement*. New York: Times Books, 1993.

Shane, Howard C., and Patti D. Albert. "Electronic Screen Media for Persons with Autism Spectrum Disorders: Results of a Survey." *Journal of Autism and Developmental Disorders* 38, no. 8 (2008): 1499–1508.

Shattuck, Paul T., Sarah Carter Narendorf, Benjamin Cooper, Paul R. Sterzing, Mary Wagner, and Julie Lounds Taylor. "Postsecondary Education and Employment among Youth with an Autism Spectrum Disorder." *Pediatrics* 129, no. 6 (2012): 1042–1049.

Shinohara, Kristen, and Jacob O. Wobbrock. "In the Shadow of Misperception: Assistive Technology Use and Social Interactions." In *Proceedings of the Annual Conference on Human Factors in Computing Systems (CHI)*, 705–714. New York: ACM Press, 2011.

Siebers, Tobin. *Disability Theory*. Ann Arbor: University of Michigan Press, 2008.

Silberman, Steve. *NeuroTribes: The Legacy of Autism and the Future of Neurodiversity*. New York: Avery, 2015.

Silverstone, Roger, and Eric Hirsch, eds. *Consuming Technologies: Media and Information in Domestic Spaces*. London: Routledge, 1992.

Silverstone, Roger, Eric Hirsch, and David Morley. Information and Communication Technologies and the Moral Economy of the Household. In *Consuming Technologies: Media and Information in Domestic Spaces*, ed. Roger Silverstone and Eric Hirsch, 15–31. London: Routledge, 1992.

Slade, Giles. *Made to Break: Technology and Obsolescence in America*. Cambridge, MA: Harvard University Press, 2006.

Small, Mario L. *Unanticipated Gains: Origins of Network Inequality in Everyday Life*. Oxford: Oxford University Press, 2010.

Smukler, David. "Unauthorized Minds: How 'Theory of Mind' Theory Misrepresents Autism." *Mental Retardation* 43, no. 1 (2005): 11–24.

Sobchack, Vivian. A Leg to Stand On: Prosthetics, Metaphor, and Materiality. In *Carnal Thoughts: Embodiment and Moving Image Culture*, 205–225. Berkeley: University of California Press, 2004.

Society of Professional Journalists. *SJP Code of Ethics*. Indianapolis, IN: Society of Professional Journalists, 2014. http://www.spj.org/ethicscode.asp.

Söderström, Sylvia. "Offline Social Ties and Online Use of Computers: A Study of Disabled Youth and Their Use of ICT Advances." *New Media and Society* 11, no. 5 (2009): 709–727.

Sosa, Anna V. "Association of the Type of Toy Used during Play with the Quantity and Quality of Parent–Infant Communication." *JAMA Pediatrics* 170, no. 2 (2016): 132–137.

Soto, Gloria. "Training Partners in AAC in Culturally Diverse Families." *Perspectives on Augmentative and Alternative Communication* 21 (2012): 144–150.

Sousa, Amy C. "From Refrigerator Mothers to Warrior-Heroes: The Cultural Identity Transformation of Mothers Raising Children with Intellectual Disabilities." *Symbolic Interaction* 34, no. 2 (2011): 220–243.

Spangler, Todd. "Kids Love Apple's iPad More than Disney, Nickelodeon, YouTube: Survey." *Variety*, October 6, 2014. http://variety.com/2014/digital/news/kids-love-apples-ipad-more-than-disney-nickelodeon-youtube-survey-1201322460/.

Spigel, Lynn. *Make Room for TV: Television and the Family Ideal in Postwar America*. Chicago: University of Chicago Press, 1992.

Spigel, Lynn. "Media Homes: Then and Now." *International Journal of Cultural Studies* 4, no. 4 (2001): 385–411.

Steiner-Adair, Catherine. *The Big Disconnect: Protecting Childhood and Family Relationships in the Digital Age*. New York: HarperCollins, 2014.

Sterne, Jonathan. *The Audible Past: Cultural Origins of Sound Reproduction*. Durham, NC: Duke University Press, 2003.

Sterne, Jonathan. "Bourdieu, Technique, and Technology." *Cultural Studies* 17, no. 3–4 (2003): 367–389.

Strauss, Anselm L., and Juliet Corbin. *Basics of Qualitative Research: Techniques and Procedures for Developing Grounded Theory*. 2nd ed. London: Sage, 1998.

Streeter, Tom. "Steve Jobs, Romantic Individualism, and the Desire for Good Capitalism." *International Journal of Communication* 9 (2015): feature 3106–3124.

Suskind, Ron. *Life, Animated: A Story of Sidekicks, Heroes, and Autism*. Glendale, CA: Kingswell, 2014.

Tacchi, Jo. Digital Engagement: Voice and Participation in Development. In *Digital Anthropology*, ed. Heather A. Horst and Daniel Miller, 225–241. New York: Berg, 2012.

Taylor, Steven J., and Robert Bogdan. *Introduction to Qualitative Research Methods*. 3rd ed. New York: Wiley, 1998.

Taylor, Sunny. "The Right Not to Work: Power and Disability." *Monthly Review* 55, no. 10 (2004). http://monthlyreview.org/2004/03/01/the-right-not-to-work-power-and-disability/.

Thoreau, Estelle. "*Ouch!* An Examination of the Self-Representation of Disabled People on the Internet." *Journal of Computer-Mediated Communication* 11, no. 2 (2006): 442–468.

Thornton, Patricia. "Communications Technology—Empowerment or Disempowerment?" *Disability, Handicap, and Society* 8, no. 4 (1993): 339–349.

Tisdall, E. Kay M. "The Challenge and Challenging of Childhood Studies? Learning from Disability Studies and Research with Disabled Children." *Children and Society* 26 (2012): 181–191.

Trainor, Audrey A. "Reexamining the Promise of Parent Participation in Special Education: An Analysis of Cultural and Social Capital." *Anthropology and Education Quarterly* 41, no. 3 (2010): 245–263.

Turkle, Sherry. Always-On/Always-On-You: The Tethered Self. In *Handbook of Mobile Communications and Social Change*, ed. James E. Katz, 121–138. Cambridge, MA: MIT Press, 2008.

Turkle, Sherry, ed. *The Inner History of Devices*. Cambridge, MA: MIT Press, 2008.

Turkle, Sherry. *Alone Together: Why We Expect More from Technology and Less from Each Other*. New York: Basic Books, 2011.

Turkle, Sherry. *Reclaiming Conversation: The Power of Talk in a Digital Age*. New York: Penguin, 2015.

Turner, Fred. "Burning Man at Google: A Cultural Infrastructure for New Media Production." *New Media and Society* 11, no. 1–2 (2009): 73–94.

Tyner, Kathleen. *Literacy in a Digital World: Teaching and Learning in the Age of Information*. Mahwah, NJ: Lawrence Erlbaum, 1998.

US Assistive Technology Act of 2004, Pub. L. No. 108–364 (2004).

US Census Bureau. *QuickFacts: California*. n.d. http://quickfacts.census.gov/qfd/states/06000.html.

US Department of Education, Institute of Education Sciences, National Center for Education Statistics. "The Condition of Education: Children and Youth with Disabilities." Last updated May 2015. http://nces.ed.gov/programs/coe/indicator_cgg.asp.

US Individuals with Disabilities Education Act Amendments of 1997, Pub. L. No. 105–17 (1997).

Valencia, Richard. *Dismantling Contemporary Deficit Thinking: Educational Thought and Practice*. London: Routledge, 2010.

Valentino-DeVries, Jennifer. "Using the iPad to Connect: Parents, Therapists Use Apple Tablet to Communicate with Special Needs Kids." *Wall Street Journal*, October 13, 2010. http://www.wsj.com/news/articles/SB10001424052748703440004575547971877769154.

Vance, Ashlee. "Insurers Fight Speech-Impairment Remedy." *New York Times*, September 14, 2009. http://www.nytimes.com/2009/09/15/technology/15speech.html.

van Leeuwen, Theo. Vox Humana: The Instrumental Representation of the Human Voice. In *Voice: Vocal Aesthetics in Digital Arts and Media*, ed. Norie Neumark, Ross Gibson, and Theo van Leeuwen, 5–15. Cambridge, MA: MIT Press, 2010.

Voskuhl, Adelheid. "Humans, Machines, and Conversations: An Ethnographic Study of the Making of Automatic Speech Recognition Technologies." *Social Studies of Science* 34, no. 3 (2004): 393–421.

Vygotsky, Lev S. *Mind in Society: The Development of Higher Psychological Processes*. Cambridge, MA: Harvard University Press, 1978.

Wajcman, Judy, and Paul K. Jones. "Border Communication: Media Sociology and STS." *Media, Culture, and Society* 34, no. 6 (2012): 673–690.

Wall Street Journal (unnamed staff reporter). "Concern Is Offering an Electronic Voice to Vocally Impaired." *Wall Street Journal*, November 2, 1977, 20.

Warner, W. Lloyd, Marchia Meeker, and Kenneth Eells. *Social Class in America: A Manual of Procedure for the Measurement of Social Status*. Oxford, UK: Science Research Associates, 1949.

Warren, Ron. "Parental Mediation of Children's Television Viewing in Low-Income Families." *Journal of Communication* 55, no. 4 (2005): 847–863.

Warschauer, Mark. *Technology and Social Inclusion: Rethinking the Digital Divide*. Cambridge, MA: MIT Press, 2003.

Warschauer, Mark, and Tina Matuchniak. "New Technology and Digital Worlds: Analyzing Evidence of Equity in Access, Use, and Outcomes." *Review of Research in Education* 34, no. 1 (2010): 179–225.

Wartella, Ellen, Victoria Rideout, Alexis Lauricella, and Sabrina Connell. *Parenting in the Age of Digital Technology: A National Survey*. Evanston, IL: Center on Media and Human Development, School of Communication, Northwestern University, 2013.

Weber, Max. *The Theory of Social and Economic Organization*. Trans. Talcott Parsons. New York: Oxford University Press, 1947.

White, Steven C., and Janet McCarty. "Reimbursement for AAC Devices." *ASHA Leader* 16 (2011): 3–7.

Wickenden, Mary. "Talking to Teenagers: Using Anthropological Methods to Explore Identity and the Lifeworlds of Young People Who Use AAC." *Communication Disorders Quarterly* 32, no. 3 (2011): 151–163.

Wiegand, Karl, and Rupal Patel. "Towards More Intelligent and Personalized AAC." Paper presented at the Supporting Children with Complex Communication Needs workshop at the annual conference on Human Factors in Computing Systems (CHI) (2014).

Wikipedia. "Stephen Hawking in Popular Culture." Last modified January 17, 2016. http://en.wikipedia.org/wiki/Stephen_Hawking_in_popular_culture.

Williams, Michael B. "Privacy and AAC." *Alternatively Speaking* 5, no. 2 (2000): 1–5.

Williams, Raymond. *Television: Technology and Cultural Form*. New York: Schocken Books, 1975.

Williamson, Bess. "Electric Moms and Quad Drivers: People with Disabilities Buying, Making, and Using Technology in Post-War America." *American Studies* 52, no. 1 (2012): 5–29.

Williamson, Bess. "Getting a Grip: Disability in American Industrial Design of the Late Twentieth Century." *Winterthur Portfolio* 46, no. 4 (2012): 213–236.

Wilson, Laurence, Joel Kaiser, and Crystal Simpson. "Final Decision Memorandum." Centers for Medicare and Medicaid Services, July 29, 2015.

Wilson, Matthew W. "Continuous Connectivity, Handheld Computers, and Mobile Spatial Knowledge." *Environment and Planning D: Society and Space* 32, no. 3 (2014): 535–555.

Wodka, Ericka L., Pamela Mathy, and Luther Kalb. "Predictors of Phrase and Fluent Speech in Children with Autism and Severe Language Delay." *Pediatrics* 131, no. 4 (2013): 1128–1134.

World Health Organization (WHO). *International Classification of Functioning, Disability, and Health (ICF)*. Geneva: WHO Press, 2002.

World Health Organization (WHO). *World Report on Disability*. Geneva: WHO Press, 2011.

Wrong, Dennis H. "The Oversocialized Conception of Man in Modern Sociology." *American Sociological Review* 26, no. 2 (1961): 183–193.

Yardi, Sarita, and Amy Bruckman. "Income, Race, and Class: Exploring Socioeconomic Differences in Family Technology Use." In *Proceedings of the Annual Conference on Human Factors in Computing Systems (CHI)*, 3041–3050. New York: ACM Press, 2012.

Zablotsky, Benjamin, Lindsey I. Black, Matthew J. Maenner, Laura A. Schieve, and Stephen J. Blumberg. Estimated Prevalence of Autism and Other Developmental Disabilities Following Questionnaire Changes in the 2014 National Health Interview Survey. In *National Health Statistics Reports No. 87*. Hyattsville, MD: National Center for Health Statistics, 2015.

Zeman, Laura D., Jayme Swanke, and Judy Doktor. "Strengths Classification of Social Relationships among Cybermothers Raising Children with Autism Spectrum Disorders." *School Community Journal* 21, no. 1 (2011): 37–51.

Zola, Irving K. "Self, Identity, and the Naming Question: Reflections on the Language of Disability." *Social Science and Medicine* 36, no. 2 (1993): 167–173.

Index

Bold page numbers refer to figures and tables.